A2 in a Week

Biology

Jim Sharpe and
Jo[illegible] Whitehead,
A[illegible]
S[illegible]

Where to find the information you need

Respiration 3
Homeostasis 9
Nervous Coordination 17
Sensitivity and Responses 25
Photosynthesis 32
Inheritance 38
Variation 49
Evolution 55
Ecology 61
Nutrient Cycles and Pollution 67
Synoptic Assessment 74
Exam Practice Questions 76
Use your Knowledge Answers 78

SUCCESS OR YOUR MONEY BACK

Letts' market leading series A2 in a Week gives you everything you need for exam success. We're so confident that they're the best revision books you can buy that if you don't make the grade we will give you your money back!

HERE'S HOW IT WORKS

Register the Letts A2 in a Week guide you buy by writing to us within 28 days of purchase with the following information:

- Name
- Address
- Postcode
- Subject of A2 in a Week book bought

Please include your till receipt

To make a **claim**, compare your results to the grades below. If any of your grades qualify for a refund, make a claim by writing to us within 28 days of getting your results, enclosing a copy of your original exam slip. If you do not register, you won't be able to make a claim after you receive your results.

CLAIM IF...

You are an A2 (A Level) student and do not get grade E or above.
You are a Scottish Higher level student and do not get a grade C or above.
This offer is not open to Scottish students taking SCE Higher Grade, or Intermediate qualifications.

Registration and claim address:
Letts Success or Your Money Back Offer, Letts Educational, Chiswick Centre, 414 Chiswick High Road, London W4 5TF

TERMS AND CONDITIONS

1. Applies to the Letts A2 in a Week series only
2. Registration of purchases must be received by Letts Educational within 28 days of the purchase date
3. Registration must be accompanied by a valid till receipt
4. All money back claims must be received by Letts Educational within 28 days of receiving exam results
5. All claims must be accompanied by a letter stating the claim and a copy of the relevant exam results slip
6. Claims will be invalid if they do not match with the original registered subjects
7. Letts Educational reserves the right to seek confirmation of the level of entry of the claimant
8. Responsibility cannot be accepted for lost, delayed or damaged applications, or applications received outside of the stated registration/claim timescales
9. Proof of posting will not be accepted as proof of delivery
10. Offer only available to A2 students studying within the UK
11. SUCCESS OR YOUR MONEY BACK is promoted by Letts Educational, Chiswick Centre, 414 Chiswick High Road, London W4 5TF
12. Registration indicates a complete acceptance of these rules
13. Illegible entries will be disqualified
14. In all matters, the decision of Letts Educational will be final and no correspondence will be entered into

Letts Educational
Chiswick Centre
414 Chiswick High Road
London W4 5TF
Tel: 020 8996 3333
Fax: 020 8742 8390
e-mail: mail@lettsed.co.uk
website: www.letts-education.com

First published 2001
Reprinted 2002
Reprinted 2004

British Library Cataloguing in Publication Data
A CIP record for this book is available from the British Library.

ISBN 1 84315 360 2

Cover design by Purple, London

Prepared by *specialist* publishing services, Milton Keynes

Printed in the UK

Letts Educational Limited is a division of Granada Learning Limited, part of Granada plc

Test your knowledge

1 Adenosine Triphosphate (ATP) is a nucleotide composed of a pentose sugar (__________), an organic base (__________) and three phosphate groups. It is produced from the reaction of ADP and __________ __________ in respiration and photosynthesis.

2 ATP is a source of __________ . It is used for the process of __________ transport, synthesis of __________ from monomers and __________ contraction.

3 A metabolic pathway is a series of chemical __________ catalysed by __________ . An __________ reaction occurs when complex molecules are built up from simpler ones.

4 Aerobic respiration requires __________ and it involves three main stages called glycolysis, the __________ __________ and the __________ __________ __________ .

5 Glycolysis involves the __________ of glucose to __________ with a net gain of __________ and reduced __________ . Initially, the glucose undergoes a __________ reaction to make it more unstable and break down into two molecules, each containing __________ carbon atoms, called glycerate 3 phosphate.

6 The Krebs cycle occurs in the __________ of a mitochondrion. Pyruvate is __________ and results in the production of __________ __________ , $NADH^+$ H^+ and more ATP.

7 The process of __________ __________ occurs on the electron transport chain on the __________ of the mitochondrion. This results in the generation of __________ .

8 The product of anaerobic respiration in muscles is called __________ __________ and in yeast, __________ and carbon dioxide are produced. Anaerobic respiration occurs in the __________ of oxygen.

9 __________ indicators are used to demonstrate oxidation reactions. __________ __________ changes from blue to colourless when hydrogen ions are accepted.

10 If a respiratory quotient of 1 is obtained, the substrate is usually a __________ . The value can be calculated by the volume of __________ __________ produced divided by the volume of oxygen __________ .

1 ribose, adenine, inorganic phosphate **2** energy, active, macromolecules, muscle **3** reactions, enzymes, anabolic **4** oxygen, Krebs cycle, electron transport chain **5** oxidation, pyruvate, ATP, NAD, phosphorylation, three **6** matrix, oxidised, carbon dioxide **7** oxidative phosphorylation, cristae, ATP **8** lactic acid, ethanol, absence **9** Redox, Methylene blue **10** carbohydrate, carbon dioxide, absorbed

If you got them all right, skip to page 8

Improve your knowledge

1

Adenosine Triphosphate is a nucleotide and consists of three parts. It contains an organic base (adenine), a pentose sugar (ribose) and three phosphate groups. It is formed in the processes of respiration and photosynthesis from adenosine diphosphate (ADP) and inorganic phosphate (P_i). When a molecule of ATP is hydrolysed, an exergonic reaction occurs. This releases about 33 kJ mol^{-1} of energy.

Key points from AS in a Week
Diffusion and active transport page 6
Role of mitochondria page 7
Pentose sugars page 13

The structure of adenosine triphosphate

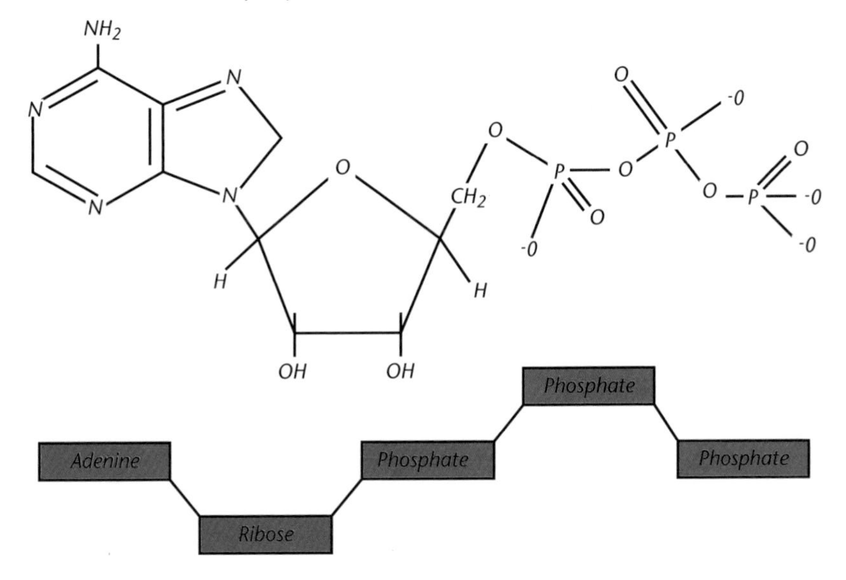

2

The energy released from ATP has the following uses:

- active transport across cell membranes (e.g. glucose uptake in the ileum)
- building macromolecules (e.g. protein from amino acids)
- movements by muscle contraction
- endocytosis and exocytosis.

Respiration

A metabolic pathway is a sequence of enzyme-controlled reactions and the molecules involved are called metabolites. Enzymes are important because they are biological catalysts, highly specific and their production is controlled by the nucleus. In anabolic reactions, complex molecules are built up from simpler ones and in catabolic reactions, complex molecules are broken down (e.g. respiration).

Aerobic respiration requires a supply of oxygen and glucose and results in the production of 38 ATP molecules, water and carbon dioxide. There are three main processes:

- glycolysis
- Krebs cycle
- electron transport chain.

5

Glycolysis takes place in the cytoplasm of the cell and involves the oxidation of glucose to the end product, pyruvate. Initially, the glucose is phosphorylated using two molecules of ATP to make it more unstable so that it readily breaks down into two molecules of glycerate 3 phosphate. A net gain of 2 ATP molecules and reduced NAD are produced.

6

Pyruvate enters the matrix of a mitochondrion and undergoes a link reaction. It is converted to acetyl (a two carbon molecule), which combines with coenzyme A to form acetyl CoA. This enters the Krebs cycle and combines with a four carbon molecule (oxaloacetate) to form a six carbon molecule (citrate). Many intermediate redox reactions occur, resulting in the release of carbon dioxide, hydrogen ions and energy in the form of heat and ATP. The hydrogen ions are removed by enzymes called dehydrogenases and reduce the coenzymes NAD and FAD.

The process of oxidative phosphorylation takes place on the cristae of a mitochondrion. This is called the electron transport chain, where electrons from the reduced coenzymes are passed to a chain of electron carriers, involving redox reactions. The energy released is used to form ATP and the final hydrogen acceptor is oxygen, which forms water.

Anaerobic respiration occurs in the absence of oxygen and therefore only glycolysis can take place. In muscles, the product is lactic acid, which is toxic and causes cramp. In yeast, the products are ethanol, carbon dioxide and NAD and it is used in the brewing industry. Overall, only 2 ATP molecules are produced, compared to 38 ATP molecules in aerobic respiration! (i.e. aerobic respiration is more efficient.)

Redox indicators are used to show when an oxidation reaction has taken place (i.e. the addition of oxygen, the removal of hydrogen ions or electrons). Methylene blue acts as an artificial hydrogen acceptor which is blue (oxidised form) and turns colourless (reduced form). Tetrazolium chloride changes from colourless (oxidised form) to pink (reduced form). These can be used as useful indicators of the speed of the action of enzymes. During the experiment, oxygen is prevented from entering to stop the oxidation of the methylene blue. A colorimeter can be used to measure the percentage of light transmitted.

10

A respiratory quotient is calculated by:

$$RQ = \frac{\text{Volume of } CO_2 \text{ produced}}{\text{Volume of } O_2 \text{ consumed}}$$

If the value is 1.0 or below, it indicates that aerobic respiration is occurring, and 1.0 or above indicates anaerobic respiration. The value also indicates the substrate being used.

RQ	Substrate
1.0	Carbohydrate
0.9	Protein
0.7	Fat

Higher respiratory quotients are obtained in animals converting carbohydrate to fat in preparation for hibernation.

Use your knowledge

Hint

1 Ruvandhi sets up the experiment below to measure the rate of anaerobic respiration in yeast with two different respiratory substrates, glucose and sucrose. Hydrogen carbonate indicator solution is used to detect carbon dioxide (an acidic gas) and changes from red to yellow on its production.

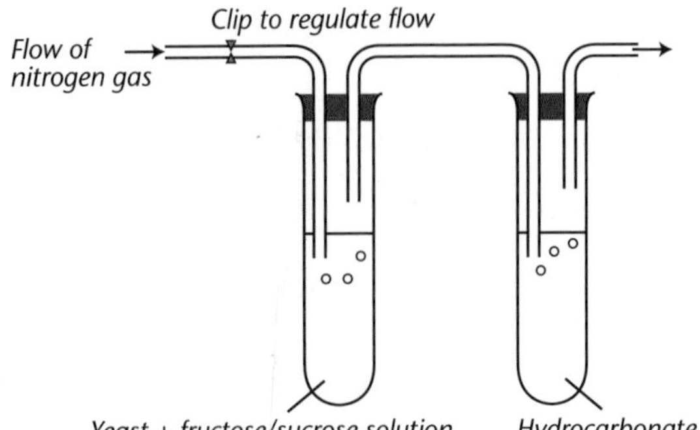

How could she use this apparatus to compare the rate of anaerobic respiration in yeast with these two sugars?

What control variables are required?

2 Cellular respiration can be divided into a series of steps: glycolysis, Krebs cycle and oxidative phosphorylation.

(a) Name a respiratory substrate that would be used in a liver cell.
(b) State in which part of the liver cell glycolysis occurs.
(c) Explain how oxidative phosphorylation produces ATP.

Test your knowledge

1. Homeostasis works through __________ feedback loops. The action of __________ is monitored by sensors and relayed to a modifying control centre.

2. Surplus amino acids cannot be stored and so are __________ in the liver. Amino groups removed enter the __________ cycle to form urea, which is removed from the blood in the __________ .

3. The basic functional units of the kidney are coiled tubes called __________ . They function both for nitrogenous excretion and __________ .

4. The __________ is a system of dense capillary networks found in the Bowman's capsule. High blood pressure is maintained by the __________ arteriole being narrower than the __________ arteriole resulting in __________ .

5. Selective reabsorption in the proximal convoluted tubule (PCT) of useful substances, e.g. __________ , involves diffusion, __________ __________ and __________ transport. PCT cells have __________ to increase surface area and mitochondria to manufacture __________ in respiration.

6. In the loop of Henle, the ascending limb is __________ to water and the descending limb is __________ . Urine concentration is controlled by __________ , which __________ permeability of the walls of the collecting duct.

7. Body temperature of __________ , e.g. insects, varies with the environment, whereas the body temperature of mammals can be decreased by physiological mechanisms such as __________ and sweating.

8. Blood temperature is monitored by the __________ . If the temperature falls, the __________ __________ centre stimulates heating mechanisms in the skin, muscles and metabolism.

9. Increased blood glucose concentrations are detected by __________ cells in the __________ ____ __________ in the pancreas, which releases the hormone __________ .

Answers

1 negative, effectors **2** deaminated, ornithine, kidney **3** nephrons, osmoregulation **4** glomerulus, efferent, afferent, ultrafiltration **5** glucose/amino acids/sodium ions, facilitated diffusion, active, microvilli, ATP **6** impermeable, permeable, ADH, increases **7** ectotherms, vasodilation **8** hypothalamus, heat gain **9** beta, islets of Langerhans, insulin.

If you got them all right, skip to page 15

Improve your knowledge

Key points from AS in a Week

Active transport, facilitated diffusion and osmosis page 6

Structure of capillaries page 48

1 Homeostasis is the maintenance of approximately constant internal body environment, e.g. body temperature. The control systems must have a sensor monitoring the environment, which relays information to a control centre that stimulates an effector to change the environment. The system works through negative feedback loops, since the action of the effector is monitored by the sensor and relayed to the control centre, which then modifies the action of the effector.

Body cells only function normally within a narrow range of conditions. Their enzymes are sensitive to temperature and pH changes, they need materials for growth and respiration, the removal of waste products and a specific composition of surrounding tissue fluid. Therefore, homeostasis provides body cells with relatively constant optimal conditions, independent of variable external conditions.

2 Excretion is the removal of the waste products of metabolism and substances in excess from the body. For example, in the liver deamination is the breakdown of surplus nitrogen-containing compounds, e.g. amino acids, which cannot be stored:

- The amino group (NH_2) is removed from the amino acid, forming ammonia (NH_3), through the addition of a hydrogen atom.
- This highly toxic ammonia is entered into the ornithine cycle (cyclical series of reactions), reacting with CO_2 to form less toxic urea, which is taken to the kidney in the blood.
- The remainder of the amino acid is used in cellular respiration.

3 Mammalian kidneys function as organs of nitrogenous excretion (filtering urea from the blood) and osmoregulation (filtering excess sodium and water). Each kidney contains a million coiled tubes (nephrons), where urine is formed. The basic structure is shown on the next page.

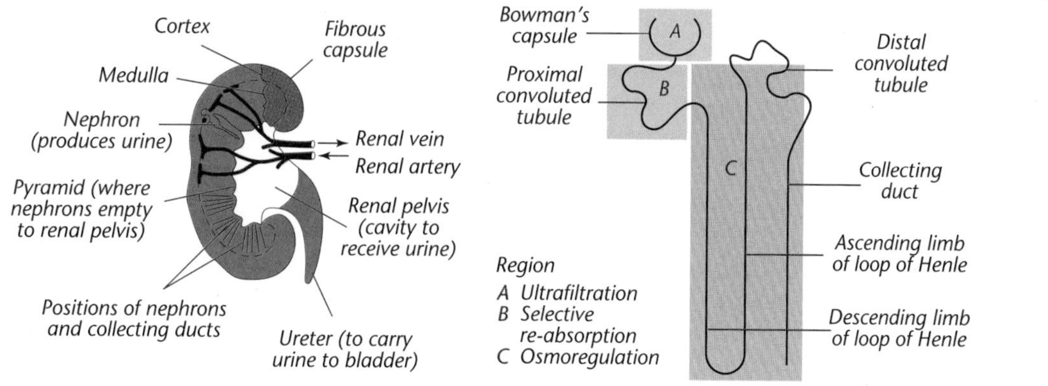

Nephrons are the functional units of the kidney

The Bowman's capsules contain dense capillary networks – glomeruli. The blood pressure is high inside these capillaries because:

- Blood in the renal artery arriving at the kidney is at high pressure.
- The efferent arteriole removing blood is narrower than the afferent arteriole supplying blood.

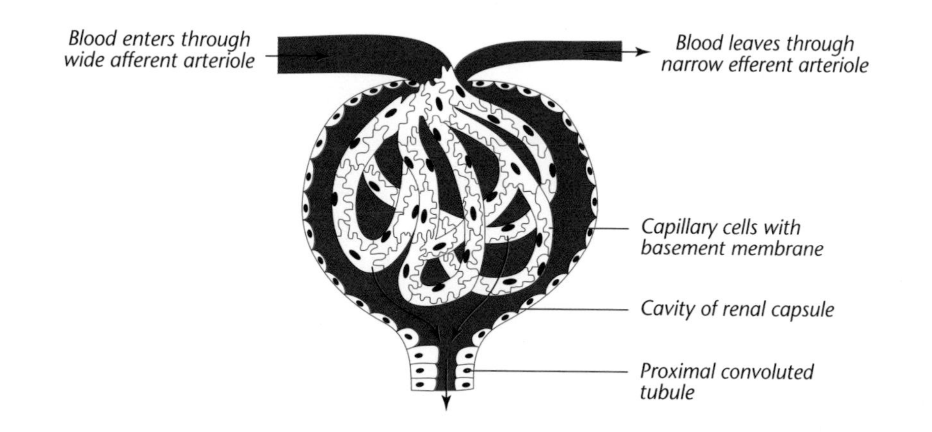

The high pressure forces small molecules, i.e. H_2O, glucose, amino acids, sodium chloride and urea from the blood, through the pores in the capillary, across the basement membrane surrounding the capillary into the nephron, forming glomerular filtrate. This process is called ultrafiltration. Larger molecules, e.g. plasma proteins, cannot pass through the filter.

5 Toxic substances in glomerular filtrate, e.g. urea, need to be removed from the body. Selective reabsorption in the proximal convoluted tubule (PCT) reabsorbs those substances present needed by the body, including glucose, amino acids and sodium ions. These diffuse from the tubule into the surrounding cells, which are adapted for reabsorption with microvilli to increase surface area and many mitochondria to manufacture ATP. Active transport then pumps these molecules into intercellular spaces. This maintains a concentration gradient, so diffusion can continue out of the tubule. From the intercellular spaces, the molecules diffuse into the blood capillaries surrounding the PCT.

Active transport requires ATP from respiration

The removal of these soluble substances results in an osmotic gradient between the filtrate and the surrounding cells. This allows reabsorption of 80–90% of the water from the tubule via osmosis back into the blood.

PCT is the main site of water reabsorption

6 The loop of Henle and the collecting duct form a system known as the countercurrent multiplier, which enables production of urine more concentrated than blood (hypertonic).

Desert mammals, e.g. gerbils, have very long loops of Henle, so produce very concentrated urine to conserve water

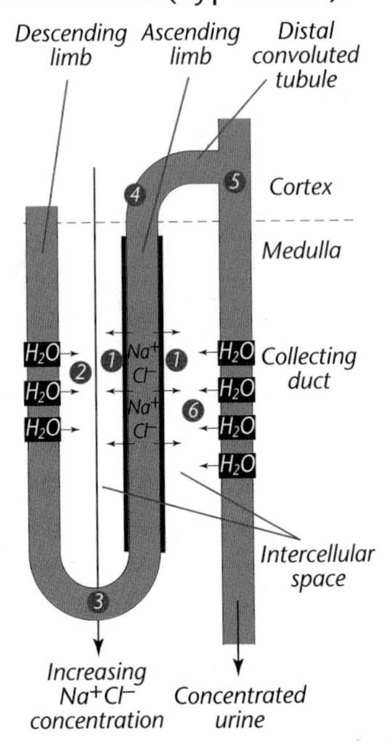

1 Na^+ and Cl^- ions are actively pumped into the intercellular space, producing a concentration gradient into the medulla. H_2O cannot follow because the ascending limb is impermeable.

2 H_2O moves out of the permeable descending limb via osmosis, increasing the filtrate concentration.

3 Filtrate at the base of the loop is very concentrated. When it enters the ascending limb, Na^+ ions are pumped out.

4 This makes the filtrate at the top of the limb dilute again, allowing more H_2O diffusion in distal convoluted tubule.
(continued opposite)

5 The filtrate drains into the collecting duct which flows back through the concentrated medulla.

6 The hormone ADH causes the walls of the collecting duct to become permeable to water, meaning water moves via osmosis out into the surrounding blood capillaries.

Temperature control is vital for enzymes to work at their optimal rate. Ectotherms, e.g. reptiles and insects, are organisms whose body temperature varies with environmental temperature. Temperature regulation occurs mainly through behavioural mechanisms, e.g. marine iguanas bask in the sun to heat up and lie on shaded, damp rocks to cool down.

Endotherms are able to colonise a wide variety of areas but have a high energy demand

Birds and mammals are endotherms. They are able to maintain their body temperature above that of the environment by behavioural and physiological mechanisms.

Cooling mechanisms in endotherms

Behavioural	Physiological
Increased convectional heat loss to the atmosphere: *flapping ears, e.g. elephants*	Increased convectional heat loss to the atmosphere: *relaxation of hair erector muscles, trapping less air around the body*
Increased heat loss via evaporation from body: *panting, e.g. dogs*	Increased heat loss via evaporation from body: *increased volume of sweat production onto skin*
Increased conduction to surroundings: *wallowing in mud, e.g. hippopotamus*	Increased radiation and convectional heat loss: *vasodilation increases flow of warm blood to the skin surface*

Make sure you can describe the warming methods of endotherms

The hypothalamus continually samples the blood, detecting small changes in temperature. When the temperature falls, nerve impulses pass from its heat gain centre, which stimulates heating mechanisms in the skin, muscles and metabolism. Conversely, the heat loss centre stimulates cooling mechanisms.

Homeostasis

9 If the blood glucose concentration falls, respiration will be slowed and cells will die, particularly in the brain. If the concentration is too high, the water potential of the blood becomes more negative and cells lose water. Therefore, mammals maintain a concentration of 80–100 mg of glucose/100 cm^3 of blood:

- Increased blood glucose concentration is detected by beta cells in the Islets of Langerhans of the pancreas, which release insulin.
- The insulin increases the uptake of glucose into cells in the liver, muscles and adipose tissue.
- Up-taken glucose is converted to glycogen, proteins or fats and broken down to CO_2 and H_2O.
- Decreased blood glucose concentration is detected by alpha cells in the Islets of Langerhans of the pancreas, which release glucagon.
- The glucagon activates enzymes that convert glycogen into glucose, which is then released into the blood. Other hormones, e.g. adrenaline, also increase blood glucose levels.

People who suffer from diabetes mellitus cannot control blood glucose levels

your knowledge

1 In an experiment, a biology student analysed and compared the composition of samples of mammalian blood plasma, glomerular filtrate and urine. Comment on the main differences you would expect to see in:

(a) glomerular filtrate and urine

(b) blood plasma and glomerular filtrate.

Hint: Think about the movement of molecules in ultrafiltration and reabsorption

2 The diagram below represents some of the mechanisms for regulating blood glucose levels in humans:

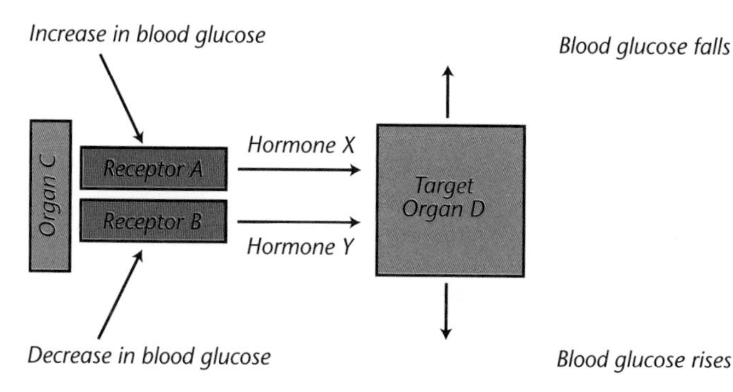

(a) (i) Identify receptor A.

(ii) Identify hormone X.

(b) Suggest a name for the target organ affected by X.

(c) Suggest why the negative feedback mechanism to regulate body temperature is faster than that regulating glucose concentrations.

Hint: The human body uses two techniques for regulating effectors

3 The graph below shows the concentration of solutes in different regions of a nephron in a human kidney.

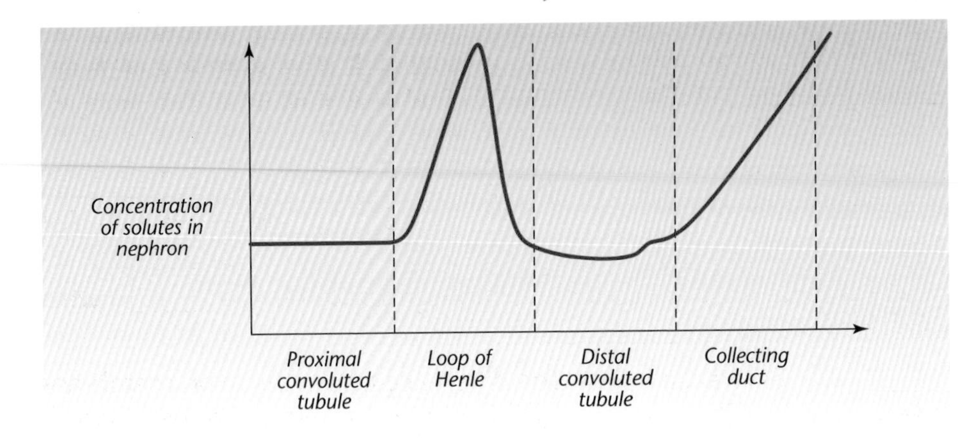

(a) Comment on the shape of the graph in terms of the filtrate passing through the loop of Henle.

(b) Explain why solute concentration does not change in the PCT.

Remember that concentrations will change due to solute and water removal

(c) Suggest whether the hormone ADH was present during the production of the urine analysed. Give a reason for your answer.

(d) State the position in the nephron at which glucose is reabsorbed.

(c) Suggest why glucose appears in the urine in experimental investigations.

Nervous Coordination

Test your knowledge

1. Hormones are produced by __________ glands and are transported in the __________________ to __________ cells. They have a _________ -lasting effect.

2. The peripheral nervous system links receptors and effectors to the _________. It can be divided into the _________ nervous system and the autonomic nervous system.

3. Sensory neurones carry impulses from receptors to the _________ , where relay neurones link them to _________ neurones.

4. A _________ difference is a charge difference across the axon membrane, by movement of Na^+/K^+ ions through membrane proteins. The membranes are impermeable to _________ ions, but permeable to _________ ions.

5. Increasing the axon diameter, ____________ the speed of an electrical impulse.

6. An impulse arriving at the synaptic knob causes an influx of _________ ions, resulting in movement of synaptic vesicles containing a ____________ .

7. Drugs can ________ neurotransmitters by amplifying their effect, known as an ____________ effect. An example of such a drug is _______ .

8. A _________ arc is an involuntary, rapid response to a specific stimulus, in order to _________ the body from damage.

9. The ______________ in the hindbrain controls ______________ and posture.

10. _________ post-synaptic potential causes hyperpolarisation. This prevents _________ , so an action potential is unlikely to occur.

Answers

1 endocrine, bloodsteam, target, long- **2** CNS, voluntary **3** CNS, motor **4** potential, Na+, K+ **5** increases **6** Ca^{2+}, neurotransmitter **7** mimic, agonistic, LSD **8** reflex, protect **9** cerebellum, balance **10** inhibitory, depolarisation

If you got them all right, skip to page 24

Nervous Coordination

Improve your knowledge

Key points from AS in a Week

Diffusion and active transport page 6

Role of mitochondria page 7

1 The differences between hormonal and nervous coordination

Hormonal Coordination	Nervous Coordination
Transmitted as chemicals in the bloodstream (i.e. takes a long time)	Transmitted as electrical impulses along neurones (i.e. takes a short time)
Hormones have an effect on target cells elsewhere in the body	Nerve impulses have a direct effect on cells
Changes metabolic activity	Causes muscles to contract or glands to secrete fluid
Long-lasting effect	Short-lived effect

2 Nervous coordination involves the detection and response to internal changes – homeostatic regulation, and the detection and response to external changes for survival, finding food and finding mates. A stimulus is any change which provokes a response. Receptors are specialised cells that detect a stimulus. Effectors are organs that bring about a response – muscles and glands.

The nervous system can be divided into two areas:

- central nervous system (CNS) – brain and spinal cord
- peripheral nervous system – nerves linking receptors and effectors to CNS.

These can be further divided into:

- voluntary nervous system – conscious control and perception
- autonomic nervous system – unconscious control (divided into sympathetic system and parasympathetic system).

3 There are three main types of neurone:

You must be able to label the different types of neurone

Sensory neurone

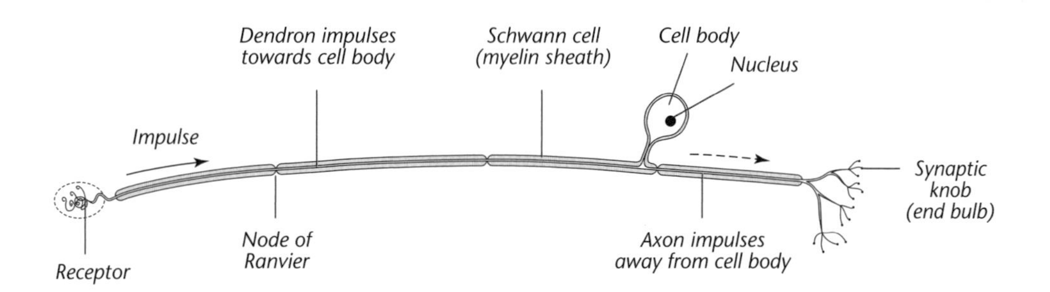

This carries impulses from receptors to the central nervous system. The cell body is found in the dorsal root ganglion of the spinal cord.

Relay neurone

This links the sensory and motor neurones, which are found in the grey matter of the brain and spinal cord.

Motor neurone

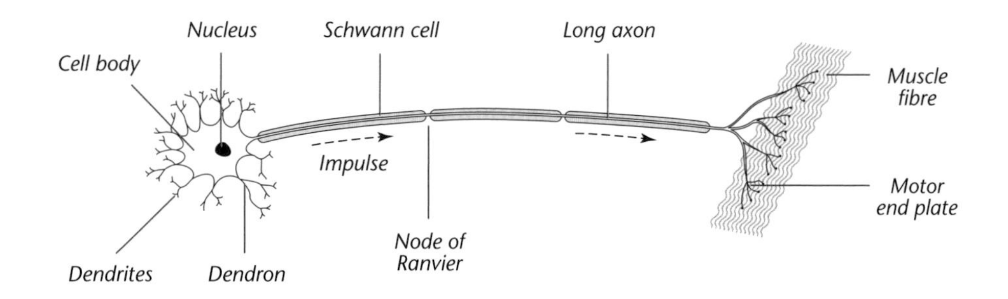

This carries impulses from the central nervous system to effectors.

4 A stimulus above the threshold (energy) level is detected by a receptor cell and causes an electrical impulse to be generated along a neurone. An action potential passes when the membrane permeability increases, causing sodium ions to flood in. This causes a positive potential difference inside the neurone – $^{+}$40 mV – and the membrane is said to be depolarised. This process is repeated along the neurone at the nodes of Ranvier.

'All or nothing' response

Differences in concentrations of ions cause differences in charge

It is repolarised by the action of the Na^+/K^+ (cation) pump in the membrane actively transporting the Na^+ ions out of the membrane and the K^+ ions into the membrane. The K^+ ions are able to diffuse into the neurone but the membrane is impermeable to the re-entry of Na^+ ions. During this time, no further impulse can be transmitted and this is called the refractory period. The movement of the sodium ions restores the resting potential and the value is usually –70 mV. The membrane is said to be polarised.

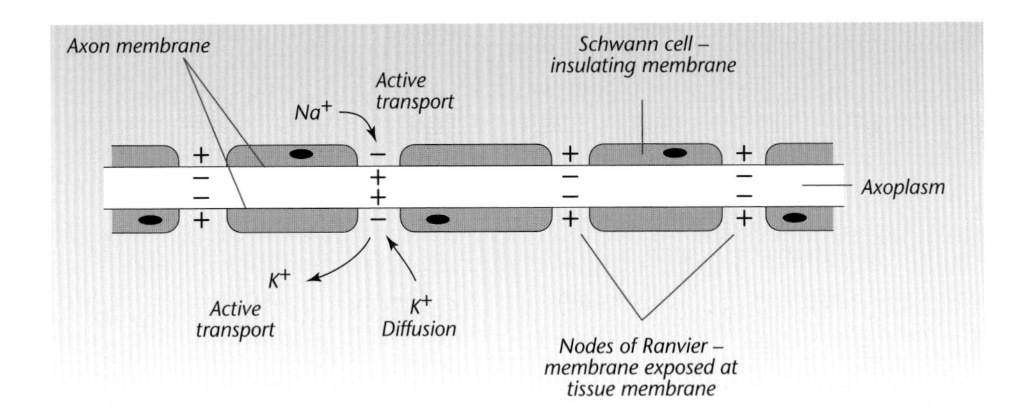

Nervous Coordination

The speed of conductance of an electrical impulse is affected by:

Factor	Speed of electrical impulse
Increasing myelination	Increases
Increasing axon diameter	Decreases
Increasing temperature	Increases

A synapse is a microscopic gap between two neurones and it is too wide for electrical impulses to cross. Impulses arriving at the synaptic knob cause an influx in calcium ions, which cause synaptic vesicles containing a neurotransmitter (chemical) to fuse with the pre-synaptic membrane. The neurotransmitter diffuses across the synaptic gap and fuses with a specific receptor on the post-synaptic membrane. This causes an action potential to be set up, resulting in sodium ions flooding in and an electrical impulse passing along the next neurone. It ensures that transmission is unidirectional.

You must be able to draw the events occurring at a synapse

Transmitter chemical	Location in nervous system	Enzyme responsible for breakdown
Acetylcholine	• voluntary (all) • autonomic (parasympathetic)	cholinesterase
Noradrenaline	• autonomic (sympathetic)	monoamine oxidase

Drugs can affect synaptic transmission in two ways.

- They can mimic natural neurotransmitters by amplifying their effect (agonistic) (e.g. LSD by causing hallucinations) or preventing the action of the enzyme which breaks down the neurotransmitter.
- Alternatively, drugs can inhibit synaptic transmission (antagonistic effect). They can destroy the neurotransmitter, inhibit its formation or prevent its action. E.g. curare blocks the receptor sites, preventing the muscle contracting, acts as a poison and causes paralysis.

Nervous Coordination

8 A reflex arc is the simplest nerve pathway and consists of three neurones.

Does not involve the conscious brain

stimulus → receptor → sensory neurone → relay neurone → motor neurone → effector

An effector is either a muscle which contracts, or a gland which secretes fluid. A reflex action is a fast, automatic response to a stimulus, which is usually concerned with the protection of the body. E.g. the pupil constricts in bright light to prevent damage to the retina.

Section through a spinal cord

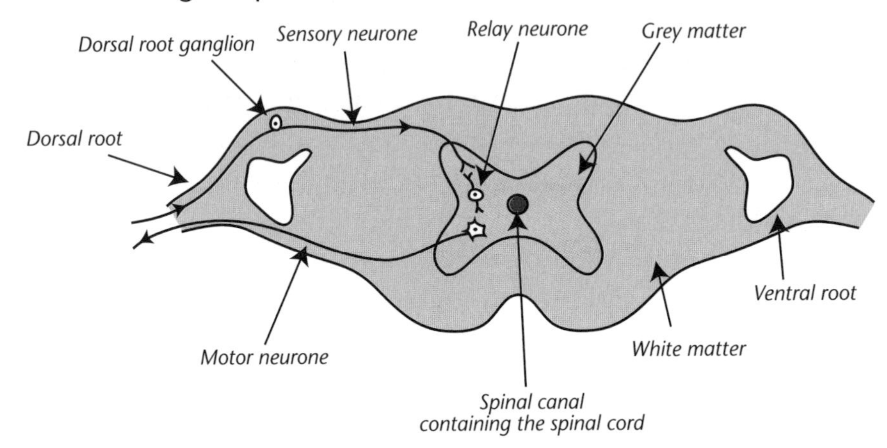

You must be able to draw the neurones accurately

9 The different parts of the brain are shown below.

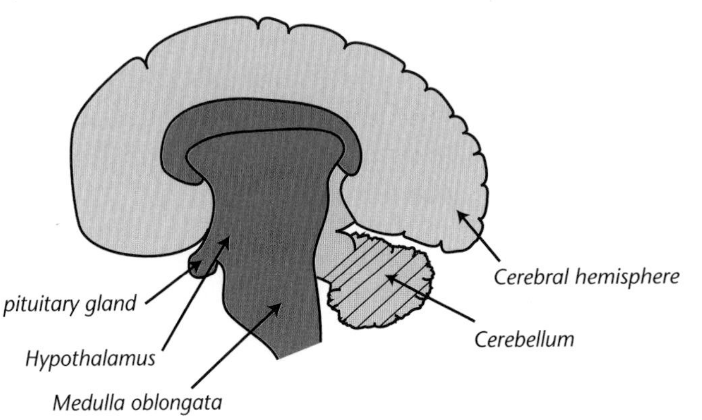

Part	Function
Medulla oblongata	Controls respiratory and heart rate
Cerebellum	Controls balance and posture
Hypothalamus	Regulates solute concentration and temperature
Cerebral hemispheres	Controls conscious decisions

Post-synaptic potentials (PSPs) produced can be in two forms:

- Excitatory (EPSP) – neurotransmitter depolarises the post-synaptic neurone, by causing an influx of Na^+ ions through membrane channel.
- Inhibitory (IPSP) – neurotransmitter opens K^+ or Cl^- channels in the post-synaptic membrane. Outflowing of K^+ or inflowing of Cl^- ions makes the membrane more negative – hyperpolarisation. This prevents depolarisation, as the membrane potential is even further from threshold. The cell is less likely to generate an action potential.

Use your knowledge

Which direction is the electrical impulse flowing in?

The diagram below shows a sensory neurone.

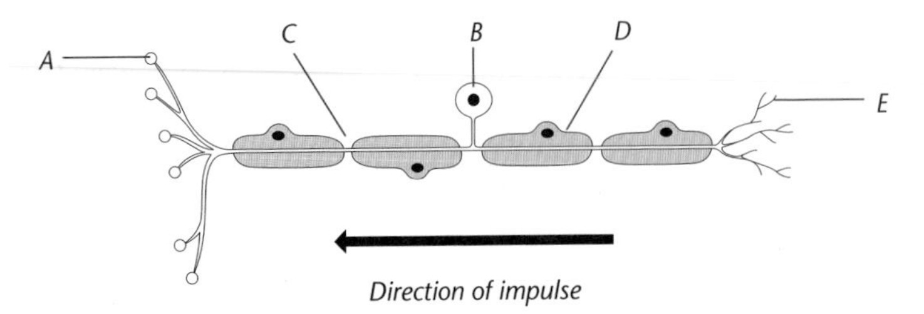

(a) Label parts A to E (5).
(b) Explain the role of the Schwann cells in the neurone.

2 (a) Draw the neurones involved with a reflex arc on the cross section of the spinal cord.

Remember the position of the neurones

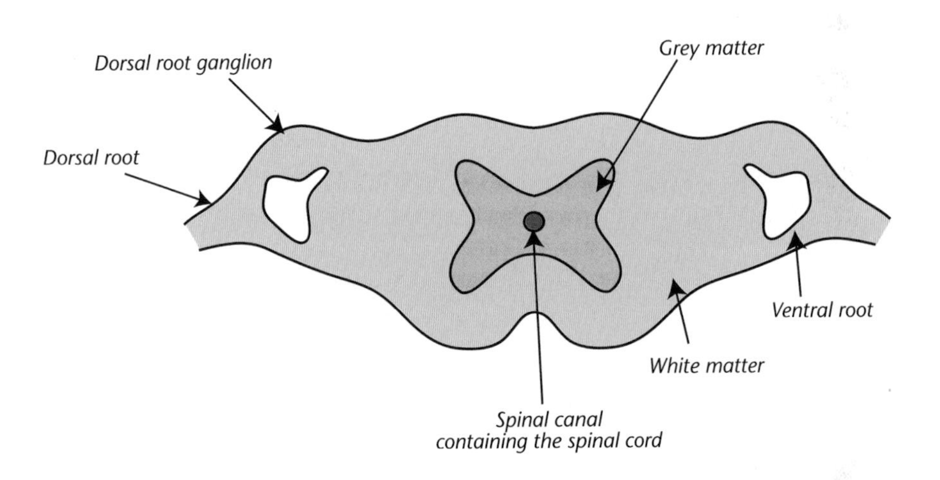

(b) What is the meaning of a 'reflex action'?
(c) Name an effector.

Sensitivity and Responses

Test your knowledge

1. In the mammalian eye, the ______ focuses images onto the retina, which contains the light-sensitive cells, except at the ______ ______ where the optic nerve leaves the eye.

2. When focusing on a near object, the ______ muscles contract, resulting in the ______ ______ slackening and the lens ______ the level of refraction.

3. Rods are found throughout the retina, except at the ______ and the blind spot. They contain the light-sensitive pigment ______ , which is sensitive to ______ light intensities.

4. The light-sensitive pigments are found in the ______ ______ of photoreceptor cells. When they are broken down by photons of light, this results in a ______ ______.

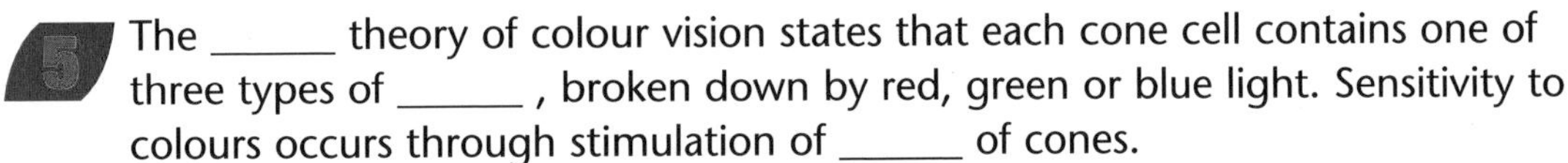

5. The ______ theory of colour vision states that each cone cell contains one of three types of ______ , broken down by red, green or blue light. Sensitivity to colours occurs through stimulation of ______ of cones.

6. ______ is a non-directional increase in movement in response to an ______ stimulus, whereas ______ is a directional response, either towards or away from a stimulus.

7. ______ corpuscles are touch or pressure receptors in the skin. Pressure causes an influx of ______ resulting in depolarisation. If the level of depolarisation exceeds a ______ level, then nervous impulses are initiated.

Answers

1 lens, blind spot **2** ciliary, suspensory ligaments, increasing **3** fovea, rhodopsin, low **4** outer segments, generator potential **5** trichromatic, iodopsin, combinations **6** Kinesis, unfavourable, taxes **7** Pacinian, Na^+, threshold

If you got them all right, skip to page 31

Sensitivity and Responses

Improve your knowledge

1 The diagram shows a section through a mammalian eye

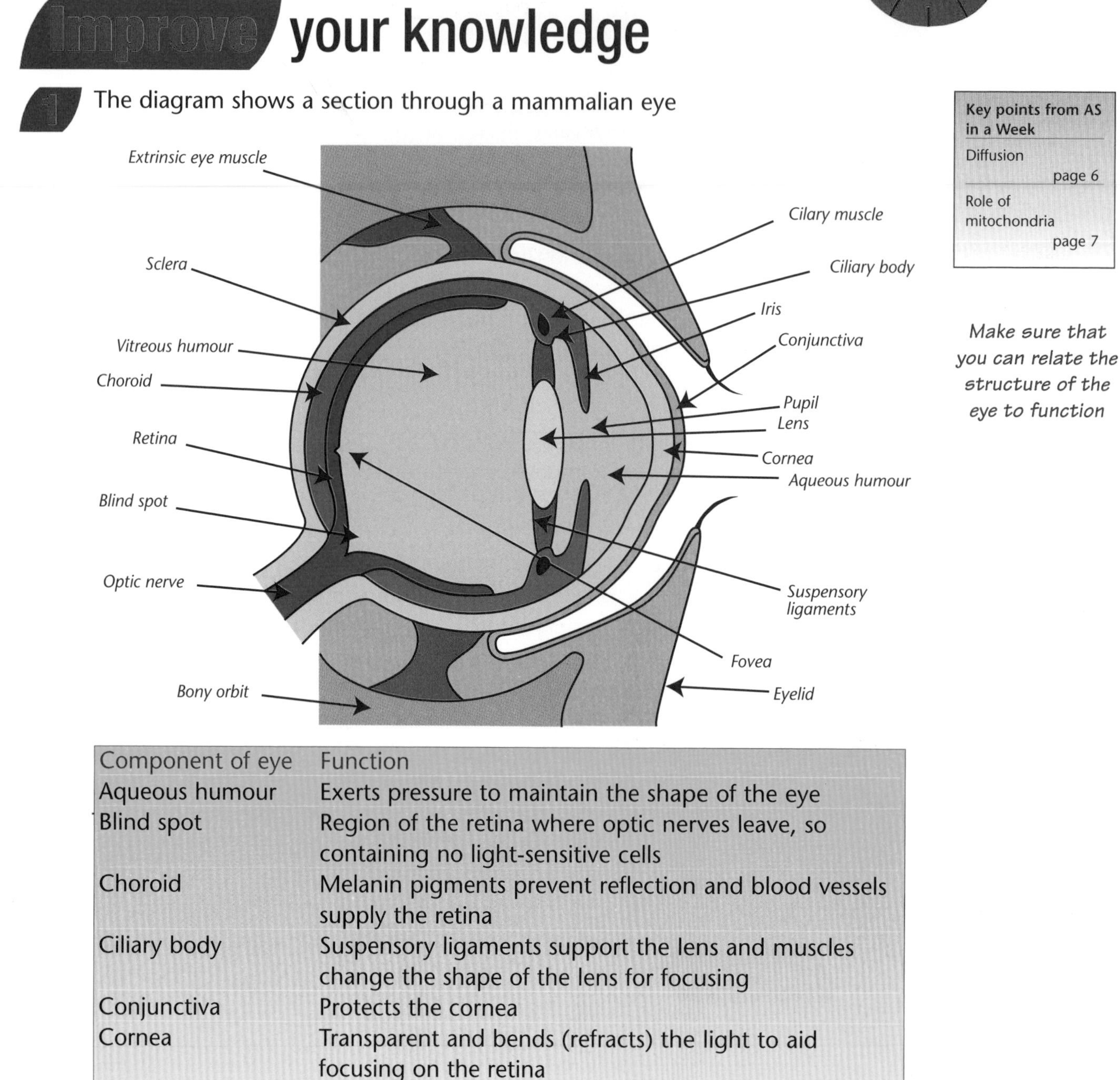

Component of eye	Function
Aqueous humour	Exerts pressure to maintain the shape of the eye
Blind spot	Region of the retina where optic nerves leave, so containing no light-sensitive cells
Choroid	Melanin pigments prevent reflection and blood vessels supply the retina
Ciliary body	Suspensory ligaments support the lens and muscles change the shape of the lens for focusing
Conjunctiva	Protects the cornea
Cornea	Transparent and bends (refracts) the light to aid focusing on the retina
Fovea	Region of the retina containing cones and rod cells

Key points from AS in a Week

Diffusion page 6

Role of mitochondria page 7

Make sure that you can relate the structure of the eye to function

Iris	Controls the size of the pupil so regulating the light entering the eye
Lens	Focuses images onto the retina
Retina	Contains the light-sensitive cells
Sclera	Outer protective layer
Vitreous humour	Exerts pressure, supporting the lens and maintaining the shape of the eye

2 Light is refracted (bent inwards) to be focused onto the cells of the retina by the conjunctiva, cornea, aqueous humour, lens and vitreous humour. Light rays from a close object are divergent and so must be refracted more than light rays from a far object, which are more parallel. Changing the shape of the lens changes the level of refraction.

Refraction is needed to produce a sharp focused image

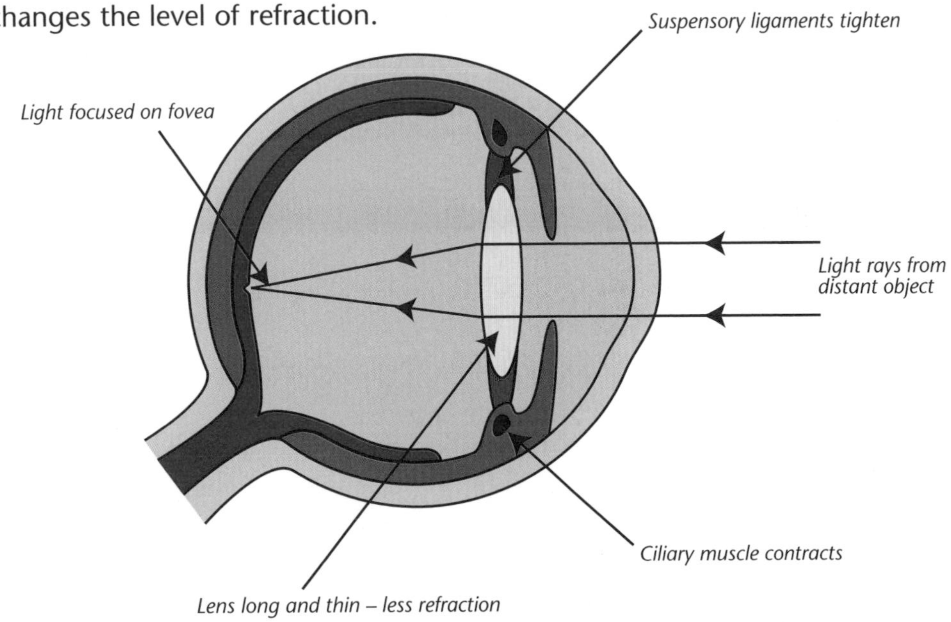

Suspensory ligaments tighten

Light focused on fovea

Light rays from distant object

Ciliary muscle contracts

Lens long and thin – less refraction

	Focusing on a near object (divergent)	Focusing on a far object (parallel)
Ciliary muscles	Contract (closing aperture around lens)	Relax (widening aperture around lens)
Suspensory ligaments	Slacken	Tighten
Lens	Return to short and fat – increasing level of refraction	Pulled long and thin – decreasing level of refraction

The process of focusing objects at different distances is called accommodation.

The arrangement of layers of cells in the retina is illustrated below.

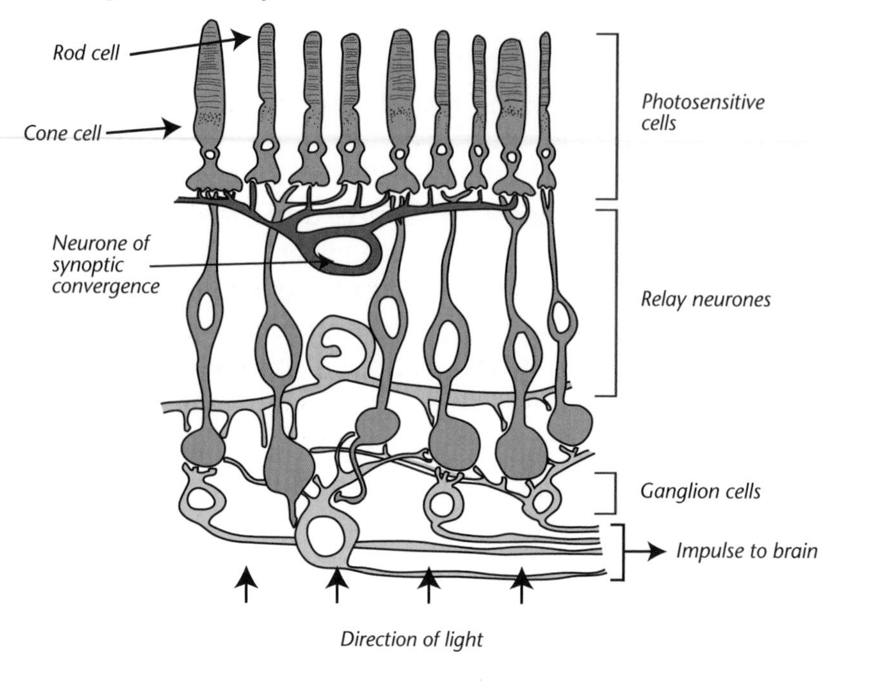

Light must pass through all the layers of the retina before striking the light-sensitive cells

Photoreceptors

The light-sensitive cells of the retina are the rods and cones.

	Rods	Cones
Location	Throughout retina except at fovea and blind spot	Occur only in the fovea
Structure of outer segment	Rod-shaped	Cone-shaped
Light-sensitive pigment	Single pigment – rhodopsin	Three pigments of iodopsin
Synapse with relay neurones	Synaptic convergence – numerous rods synapse with one relay neurone	One cone to one relay neurone

Impulses generated from the rods and cones pass via the relay neurones and the ganglion cells to the sensory neurones of the optic nerve. They are received by the brain's primary visual area in the occipital lobe of the cerebral hemispheres.

Light-sensitive pigments in the outer segments of photoreceptor cells are broken down by photons of light. For example, rhodopsin is broken down to opsin and retinal. This results in a generator potential within the rod cells, which when large enough causes an impulse along the optic nerve to the brain.

Rods are sensitive to low light intensities, because rhodopsin is more easily broken down by light than iodopsin. This sensitivity is increased by synaptic convergence, because the cumulative effect (summation) of a small generator potential in numerous rod cells is sufficient to produce an impulse in the relay neurone they all synapse with. Before it can be stimulated again, rhodopsin must be resynthesised, using ATP.

Synaptic convergence is shown on page 28

The grouping of synaptic convergence means that rods have a poor ability to distinguish between two close objects (low visual acuity). Cones have a high visual acuity because one cone synapses with one relay neurone, so each generates an individual impulse.

5

Colour vision is via the cones. Each cone cell contains one of three types of iodopsin, broken down by red, green or blue light, e.g. red light is perceived through stimulation of the red cones only. Sensitivity to other light colours occurs through stimulation of combinations of cones, e.g. perception of orange/yellow occurs through stimulation of red and green cones. This is known as the trichromatic theory of colour vision.

6

Reflexes are automatic responses to stimuli which involve a minimum of central processing. They often have a protective function or maintain organisms within favourable conditions.

Kinesis is a non-directional increase in movement in response to an unfavourable stimulus. For example, woodlice show increased speed and changes in direction of movement as conditions become drier.

A stimulus can be considered as any environmental change

Sensitivity and Responses

Taxes is a directional response, either towards (positive) or away from (negative) a stimulus. For example, many photosynthetic protoctists are positively phototactic, swimming towards light.

All sensory cells or receptors only respond to one particular type of stimulus. For example, Pacinian corpuscles are touch or pressure receptors in the skin, similar in structure to sensory neurones. The sensitive region is one modified end (or terminal), where changes in pressure are transmitted to the core. Change in shape of stretch-mediated sodium channels causes increased permeability to Na^+ ions. This influx of Na^+ causes depolarisation, producing a generator potential. If the level of depolarisation exceeds a threshold level then nervous impulses are initiated.

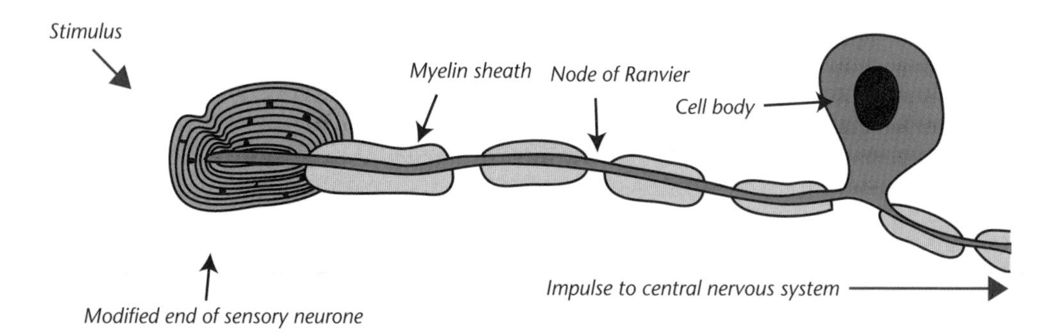

Use your knowledge

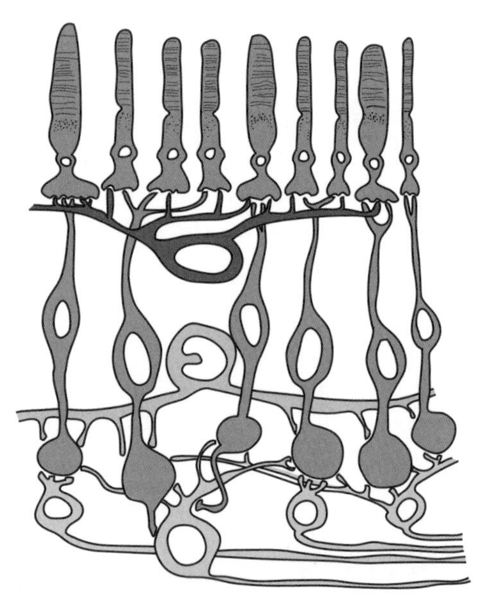

1

(a) State the names of the two types of photoreceptor present.
(b) Explain how the two types of light-sensitive cells differ in their sensitivity.
(c) Briefly summarise the trichromatic theory of colour vision.
(d) Describe how the eye focuses light from a near object onto the retina.
(e) Describe how the eye regulates light reaching the retina.

Learn precisely the differences between the two types of photoreceptors

2

An experiment was conducted on the behaviour of woodlice in different humidities. Individual organisms were transferred from a humid tank to a glass bowl in which the air was chemically dried.

(a) Describe the characteristics of the response that you would expect to observe in the woodlice.
(b) Suggest the advantage of the behaviour you have described in (a).
(c) Differentiate between the following terms:
  (i) taxes
  (ii) kinesis.

Woodlice live in damp, dark habitats, e.g. in rotting wood

Test your knowledge

1. Photosynthesis is an example of __________ nutrition involving the reduction of CO_2 to organic carbohydrate using __________ energy.

2. __________ are the organs of photosynthesis. They contain __________ mesophyll cells which are closely packed and contain a large number of chloroplasts.

3. Photosynthetic pigments are found on the chloroplast's __________ membranes. Pigments are __________ , which absorbs red and blue-violet light, and accessory pigments the __________.

4. The light-dependent reactions convert __________ in sunlight into chemical energy. __________ is the production of ATP and $NADPH_2$, with electrons from PS2 passing to PS1, whereas non-cyclic photophosphorylation only produces __________.

5. In the light-independent reactions, ribulose bisphosphate fixes __________. Triose phosphate produced is used to regenerate __________ and assimilate organic molecules.

6. An increase in temperature ______________ the rate of photosynthesis up to an optimum. After the optimum, the rate decreases because the ______________ in the Calvin cycle have become denatured.

7. When the rate of respiration is equal to the rate of photosynthesis, there is no exchange of oxygen and carbon dioxide. This is called the ______________ ______________.

8. Nitrate ions are taken up by ______________ ______________ into root hair cells and are needed by plants for the manufacture of __________, RNA and ______________ ______________.

Answers

1 autotrophic, light **2** Leaves, palisade **3** thylakoid, chlorophyll, carotenoids
4 energy, Photophosphorylation, $NADPH_2$ **5** CO_2, RuBP **6** increases, enzymes
7 compensation point **8** active transport, DNA, amino acids

If you got them all right, skip to page 37

Photosynthesis

Improve your knowledge

1 Autotrophic nutrition is the use of light energy or chemical energy to manufacture sugars, fats and proteins needed for cellular metabolism. Photosynthesis is an example and involves the conversion of inorganic carbon dioxide to organic carbohydrate using light energy. The overall chemical equation for photosynthesis is:

$$6\,CO_2 + 6\,H_2O \xrightarrow[\text{by chlorophyll}]{\text{light energy trapped}} C_6H_{12}O_6 + 6\,O_2$$

Plants are adapted to photosynthesise efficiently by maximising absorption of sunlight and maintaining a supply of raw materials.

Key points from AS in a Week
Role of chloroplast page 8
Organic molecules pages 12–14
Properties of enzymes pages 21–24

2

Structure	Function	Adaptation
Leaf	Organ of photosynthesis	• Large, flat surface area (lamina) • Supported by turgor pressure and veins • Supply of water and minerals via veins and midrib
Palisade mesophyll cells	Main photosynthetic cells	• Large number of chloroplasts in cytoplasm • Closely packed in rows under leaf surface
Chloroplast	Site of photosynthesis	• Thylakoid membranes and grana increasing internal surface area – site of pigment molecules • Stroma (gel-like matrix) containing enzymes for light-independent stage • Contain photosynthetic pigment

Photosynthesis

3 Plants use photosynthetic pigments found on the thylakoid membranes of chloroplasts to absorb light energy. The main pigments are chlorophylls a and b, which absorb red and blue–violet light. Carotenoids act as accessory pigments absorbing in the blue–violet wavelength range that chlorophylls do not effectively absorb and protect chlorophyll from excess light.

These photosynthetic pigments occur grouped together as two forms: photosystem 1 (PS1) and photosystem 2 (PS2). The collection of pigments grouped together maximises efficiency of light entrapment.

4 Photosynthesis occurs in two stages. First, the light-dependent reactions convert solar energy in sunlight into chemical energy, as ATP and reduced NADP ($NADPH_2$).

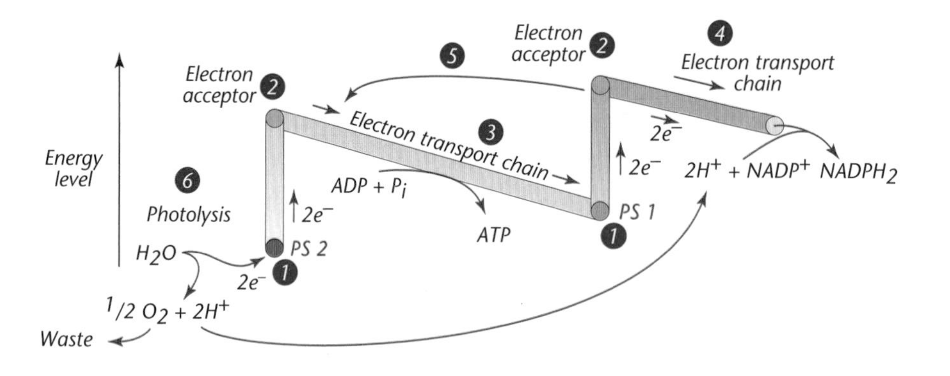

Solar energy is transferred to the electrons, exciting them to higher energy levels

1. Absorption of visible light energy by photosynthetic pigments in PS1 and PS2 causes electrons to achieve an excited state.
2. These electrons are picked up at high energy levels by electron acceptors.
3. Electrons from PS2 pass through a series of electron carriers of progressively lower energy levels. The energy released is used to form ATP – photophosphorylation. The electrons are finally acquired by positively charged and unstable PS1.
4. Electrons from PS1 can be passed along a series of electron carriers and with protons from H_2O reduce NADP to $NADPH_2$ – non-cyclic photophosphorylation.

PS1 and PS2 must replace excited state electrons to become stable

5 The electrons can alternatively be recycled to PS1 and produce ATP – cyclic photophosphorylation.
6 PS2 obtains replacement electrons from photolysis of H_2O (splitting H_2O using light energy). This also releases protons (H^+) and O_2.

5 Second, the **light-independent reaction** fixes CO_2 producing organic molecules using the chemical energy from the light-dependent reaction.

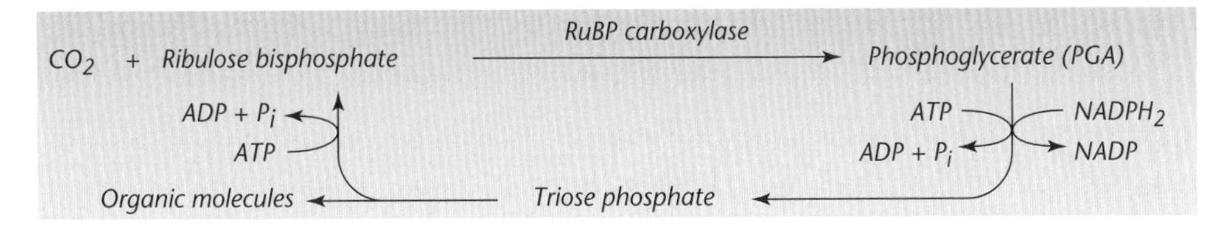

- Five-carbon RuBP accepts CO_2, catalysed by the enzyme RuBP carboxylase.
- The unstable six-carbon molecule breaks down to two molecules of three-carbon PGA.
- PGA is reduced to triose phosphate using the NADPH2 and ATP (from the light-dependent reaction).
- Triose phosphate is used both to regenerate RuBP (requiring ATP) and assimilate organic molecules, mainly glucose and starch.

6 The rate of photosynthesis can be determined by different environmental factors.

Increasing factor	Effect on rate of photosynthesis
Light intensity	Increases
CO_2 concentration	Increases
Temperature	Increases (up to an optimum, after this temperature enzymes become denatured)

The rate of photosynthesis is limited to the factor that is at its minimum value and this is called a limiting factor. This information can be used to increase crop yield.

Photosynthesis

A graph to show the effect of light intensity on the rate of photosynthesis

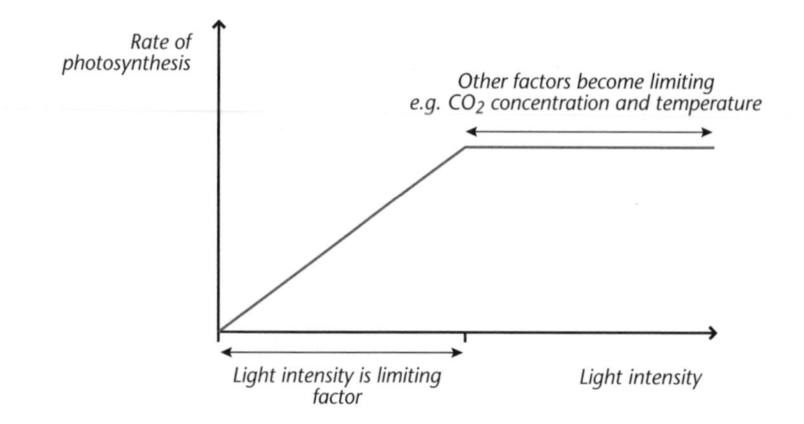

7 The compensation point occurs when the rate of respiration equals the rate of photosynthesis and this results in no exchange of oxygen or carbon dioxide – i.e. the oxygen evolved by the process of photosynthesis is used for the process of respiration. This occurs at dawn and dusk.

8 Mineral ions are taken up in small quantities by the root hair cells by active transport.

Mineral Ion	Function	Deficiency Symptoms
Nitrate	DNA, RNA and amino acids	Stunted growth
Phosphate	DNA, RNA and phospholipids	Stunted growth
Magnesium	Chlorophyll manufacture	Chlorosis

Photosynthesis

your knowledge

Remember the structure of the chloroplast

1
(a) State where in the chloroplast the light-dependent reaction takes place.
(b) In the light-dependent reaction molecules are broken down, producing oxygen, electrons and hydrogen ions.
 (i) What is the name given to the process of breaking water down?
 (ii) What is the fate of the electrons produced?
 (iii) What happens to the hydrogen ions?

Three plants of the same species, size and age (A, B and C) were placed in varying conditions of light and temperature.

Which wavelengths of light are used in photosynthesis?

(a) In the first experiment all the plants were kept in the same conditions, except: plant A received white light, plant B received red light, and plant C received green light.
 (i) State and explain which plant will have the fastest rate of photosynthesis.
 (ii) State and explain which plant will have the slowest rate of photosynthesis.
(b) State two factors other than light that will affect the rate of photosynthesis.

The production of gametes for sexual reproduction occurs through ______. The four resultant daughter nuclei are ______ , so that the fusion of gametes at fertilisation restores the ______ number.

During prophase I of meiosis, homologous chromosomes pair up to form ______. These are linked at points along their length called ______.

Anaphase II is the separation of the daughter ______ by pulling apart centromeres, producing ______ haploid cells.

Meiosis increases genetic variation through ______ and the ______ arrangement of chromosomes during metaphase I.

Mendel identified ______ characteristics that one parent may show, which do not appear in the F_1 generation, but then ______ in the F_2 generation. He concluded that individuals contain ______ 'factors' for a characteristic.

6 The law of ______ states that pairs of factors determine the characteristics of organisms, but only one factor of the pair enters a gamete. These factors are variations of a gene, called ______.

Characteristics are controlled by ______ of alleles, which are found at similar loci on ______ chromosomes.

8 A ______ cross is used to determine the genotype of individuals showing the ______ phenotype. This individual is crossed with a ______ individual.

The law of ______ ______ was Mendel's second law. This states that either member of a pair of alleles can combine with either of a second pair during ______ formation.

Inheritance

10 ______ for different characteristics behave independently when carried on ______ pairs of homologous chromosomes. This leads to independent assortment of characteristics.

11 Gametes contain one sex chromosome, so males are ______ (producing X and Y sperm), whereas females are ______.

12 Human blood group is controlled by ______ alleles. I^A and I^B show ______ over each other with I^O ______ to both.

Answers

1 meiosis, haploid, diploid **2** bivalents, chiasmata **3** chromatids, four **4** chiasmata, independent **5** recessive, reappear, two **6** segregation, alleles **7** pairs, homologous **8** back, dominant, recessive **9** independent assortment, gamete **10** Alleles, separate **11** heterogametic, homogametic **12** multiple, co-dominance, recessive

If you got them all right, skip to page 47

Improve your knowledge

1 Genetics is the study of inheritance, i.e. how characteristics are passed from one generation to another, via reproduction. The production of gametes for sexual reproduction occurs through meiosis. This is the division of a nucleus following chromosome replication (to produce daughter nuclei), involving two successive divisions. The four resultant daughter nuclei are haploid (contain half as many chromosomes as the diploid parental cells).

Key points from AS in a Week
Nuclear division pages 33–39
Reproduction pages 60–65

In humans, normal cells contain 46 chromosomes (23 homologous pairs). Gametes contain 23 unpaired chromosomes, so that the fusion of gametes at fertilisation restores the diploid number. This conserves the number of chromosomes, since diploid gametes would double the number of chromosomes every generation.

Diploid cells contain a complete set of chromosomes

2 The first stage of meiosis is prophase I. Chromosomes (visible as two daughter chromatids) condense and homologous pairs come together, forming bivalents. These bivalents are linked at points along their length (chiasmata).

- Metaphase I begins when the nuclear envelope breaks down and the bivalents line up on microtubules, called spindle fibres, at regions called kinetochores along the equator of the cell.
- The homologous chromosomes separate during anaphase I, with one member of each pair moving to the poles.
- Telophase I is the final stage of the first division and is the reverse of prophase I. Cytokinesis (cytoplasmic division) occurs, with filaments of the proteins actin and myosin contracting to produce cleavage furrow at the equator of the cell. This produces two cells containing one member of each homologous pairing. Each chromosome is still made up of two daughter chromatids.

Division 1 separates homologous chromosomes

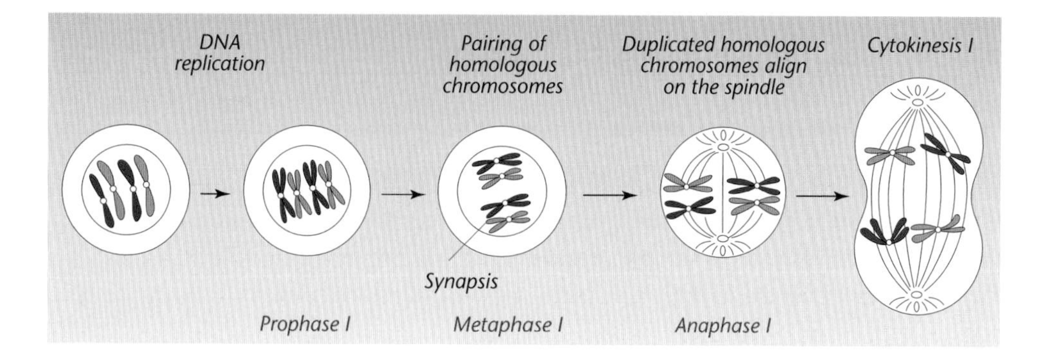

3 The second meiotic division occurs in both cells produced during division 1. In prophase II chromosomes condense and become visible again. The chromosomes then become attached to the spindle fibres and line up individually at the equator of the cells in metaphase II. Anaphase II is the separation of the daughter chromatids by pulling apart the centromeres. Finally, telophase II produces four genetically different, haploid cells, following cytokinesis.

Division 2 separates daughter chromatids

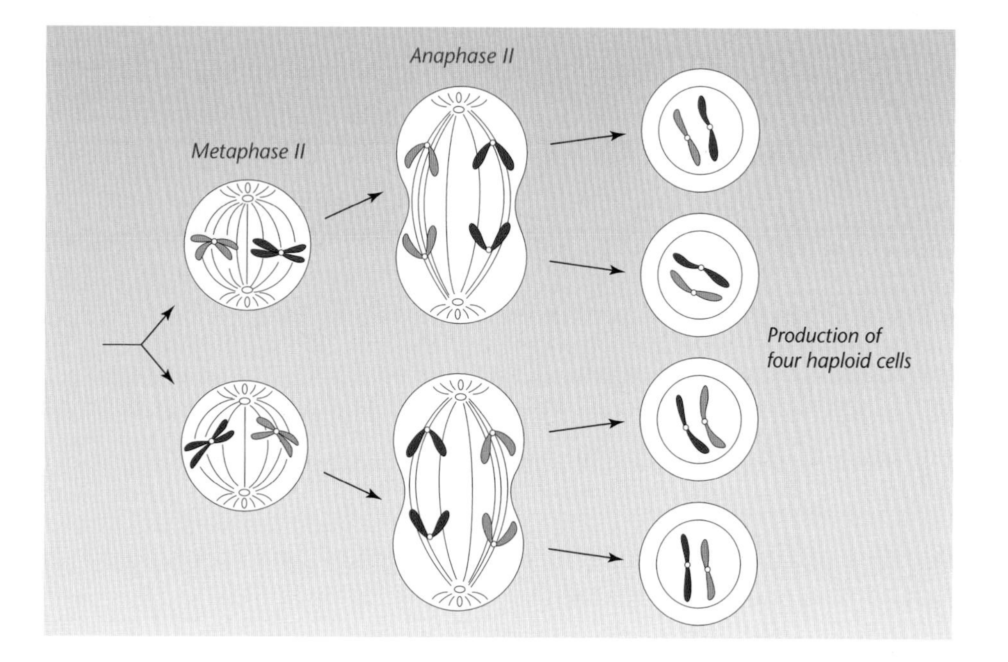

Meiosis results in genetic variation within a population:

- The independence of chromosome arrangement on the spindle at metaphase I, results in the number of possible chromosome combinations in a gamete being very large.
- Chiasmata linkage results in crossing over of genetic material between chromosomes. Chiasmata are random and produce genetic recombinants (new combinations of genes).

Gametes have a wide genetic diversity

Mendelian genetics are patterns of inheritance worked out by Gregor Mendel cross-breeding garden peas in the nineteenth century. The simplest crosses are between two individuals which differ in one characteristic – monohybrid crosses.

Mendel observed that when two individuals differing in one single characteristic were crossed, the first generation (F_1) all resemble one of the parents. If any F_1 individual is then self-pollinated, the second generation (F_2) may have either of the original parents' characteristics, in an approximately 3:1 F_2 ratio.

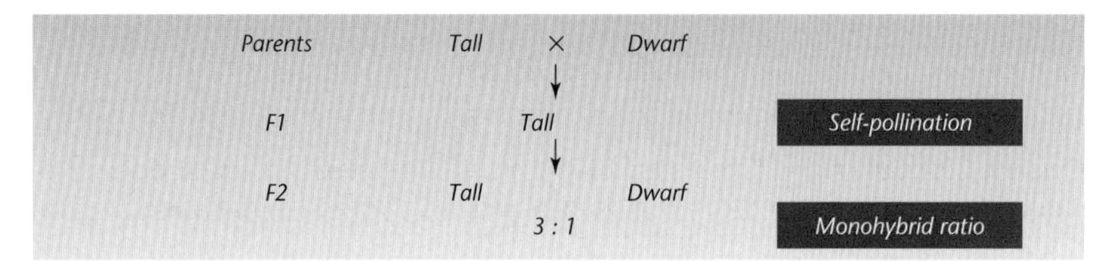

Mendel made the following deductions from this cross:

- Parental characteristics are not blended, as there are no medium-sized plants produced.
- Dwarf characteristic is recessive and tall is dominant, because no dwarf plants occurred in the F_1 generation.
- The reappearance of the dwarf plant in F_2 means that the F_1 plants, which are tall, must contain one 'factor' for tallness and one for dwarfness.

This led to Mendel's first law – the law of segregation. The characteristics of organisms are determined by pairs of 'factors', with only one factor of the pair entering a gamete. These 'factors' are known as alleles, defined as alternative forms of a gene, i.e. the gene for height can be tall or dwarf.

Let T = Gene for tall (dominant) and t = gene for dwarf (recessive)

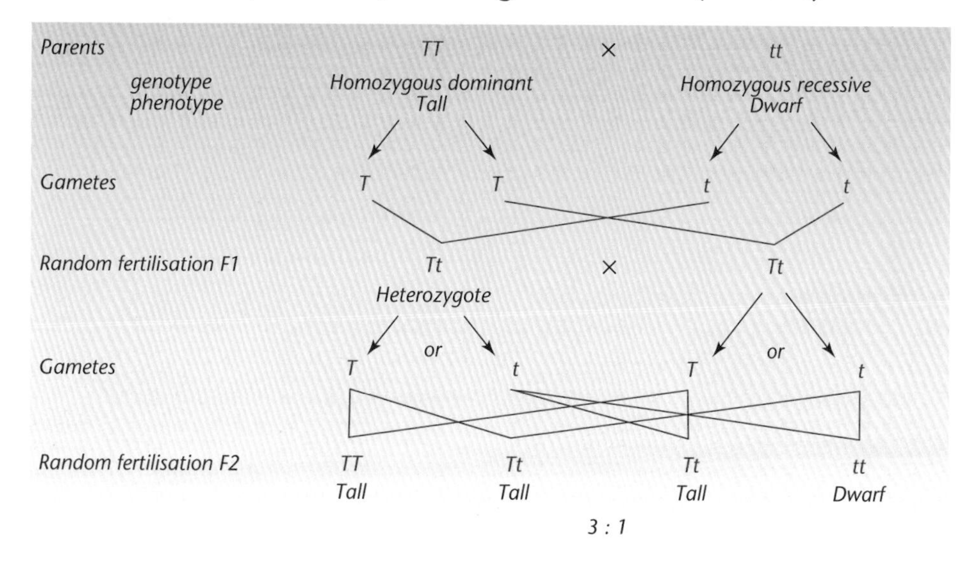

Diploid individuals contain two alleles controlling most characteristics, with one allele inherited from each parent. These pairs of alleles are found at the same position, or locus, on homologous chromosomes.

Because homologous chromosomes are separated at anaphase 1 of meiosis, during gametogenesis pairs of alleles are separated. The consequence of this is:

Gametogenesis is the production of gametes

- Gametes contain only one member of each pair of alleles.
- Fertilisation combining gametes results in new pairs of alleles forming.

A back cross or a test cross is used to determine the genotype (alleles present) in individuals with a dominant phenotype (physical appearance), i.e. a tall individual could be genotype TT or Tt. The dominant phenotype individual is crossed with a recessive individual, whose genotype must be tt.

Presence of a dominant allele always causes an individual to show dominant characteristics

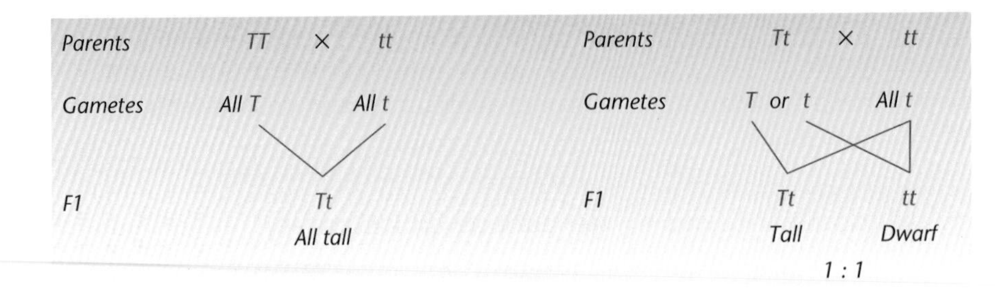

If the F_1 generation are:

- all tall – dominant parent's genotype is TT (homozygous)
- ratio 1:1 tall:dwarf – dominant parent's genotype is Tt (heterozygous).

9 Dihybrid crosses study the inheritance of two pairs of characteristics. In the cross below, the parents are homozygous for both characteristics.

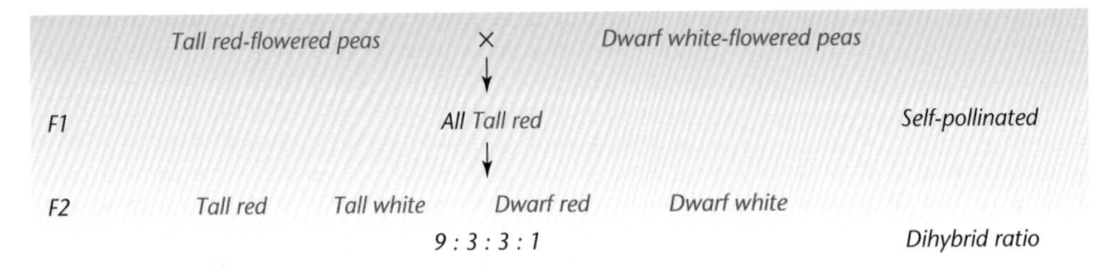

Mendel made the following deductions from this cross:

- Since the F_1 are all tall and red, these characteristics are dominant over dwarf and white respectively.
- In the F_2 generation, two new combinations appear: tall and white, and dwarf and red. Therefore, the alleles for height and colour can be separated.
- The F_2 ratio for tall:dwarf and red:white are both 3:1. Therefore, the alleles are behaving independently.

This led to Mendel's second law – law of independent assortment. Either member of a pair of alleles can combine with either member of a second pair in gametogenesis. In the dihybrid cross shown, either of the alleles for height can be in a gamete with either for flower colour. Therefore, the F_1 generation has four possible gametes.

Characteristics are inherited separately, if alleles are on separate chromosomes

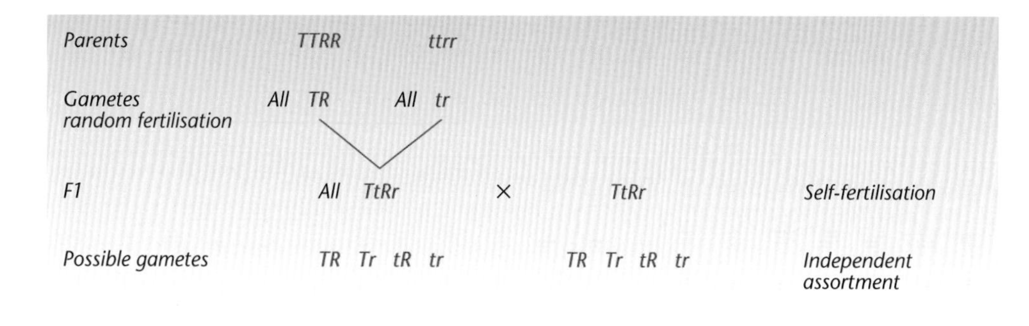

A punnett square is used to show possible offspring from the random fertilisation between these possible gametes:

	Female					
	Gametes	TR	Tr	tR	tr	
Male	TR	TTRR	TTRr	TtRR	TtRr	possible offspring genotype
	Tr	TTRr	TTrr	TtRr	Ttrr	
	tR	TtRR	TtRr	ttRR	ttRr	
	tr	TtRr	Ttrr	ttRr	ttrr	

F_2 phenotypes:
9 – tall, red. 3 – dwarf, red. 3 – tall, white. 1 – dwarf, white.

Independent assortment (alleles for different characteristics behaving independently) occurs when they are carried on separate pairs of homologous chromosomes. The homologous pairs line up independently on the equator during metaphase 1 of meiosis. It is a matter of chance which combination of alleles move to a pole together and go into a gamete.

A back cross or a test cross is used to determine the genotype of an individual with dominant characteristics. For example, a tall, red plant can be TTRR, TtRR, TTRr or TtRr (see overleaf).

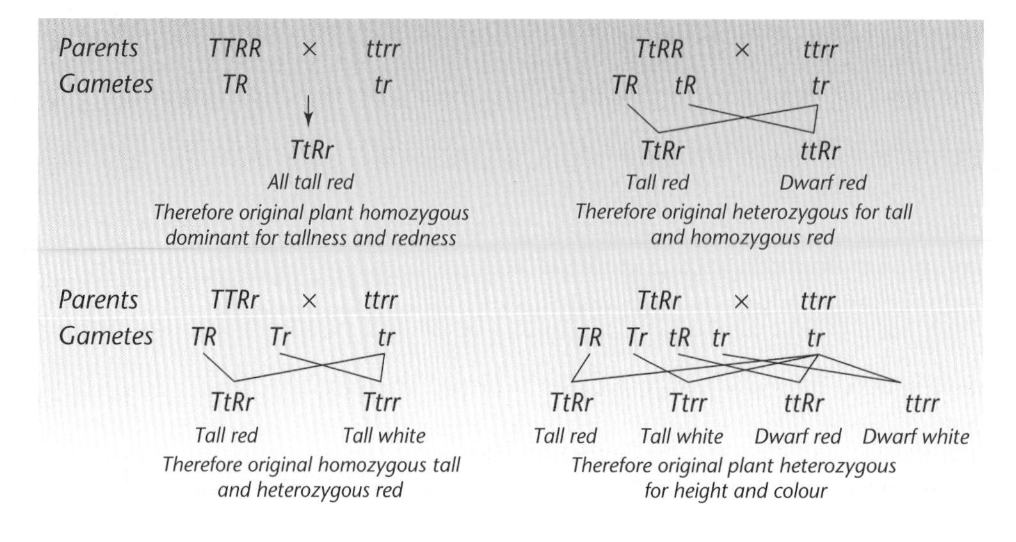

11 The sex of an individual is determined by a pair of sex chromosomes – X and Y. The X chromosome is much larger than the Y and since the Y is much smaller it carries very few genes. In mammals (including humans), females are XX (homozygous) and males are XY (heterozygous). However, in some organisms, e.g. butterflies, males are XX and females are XY.

In meiosis, sex chromosomes pair up and are separated to opposite poles of the cell. Therefore, since gametes only contain one sex chromosome, males are heterogametic (producing X and Y sperm) and females homogametic (producing X ova).

In exam questions involving sex of organisms, check the genotype very carefully

12 Human blood group is controlled by multiple alleles. There are three alternative allele forms: I^A, I^B and I^O, although each individual still only posseses two of these alleles. I^A and I^B show incomplete dominance over each other (co-dominance), with I^O recessive to both.

Phenotype (blood group)	A	B	O	AB
Genotype	$I^A I^A$ or $I^A I^O$	$I^B I^B$ or $I^B I^O$	$I^O I^O$	$I^A I^B$

Use your knowledge

Hint

1 The gene for petal colour in a flowering plant has two alleles: W = white and w = pink. The gene for leaf shape is controlled by a pair of alleles: R = round and r = oval. The two pairs are carried on separate homologous chromosomes.

(a) (i) Define the term homologous chromosome.
(ii) Distinguish between the term gene and allele. (2)

(b) State the phenotype of the following genotypes (2):
(i) WWrr
(ii) WwRr.

(c) State the possible genotypes of the following plants:
(i) pink flower with round leaves (2)
(ii) pink flower oval leaves. (1)

Gametes have one allele for each characteristic; diploid cells have two

2 Peach trees show variation in their height (tall or dwarf) and leaf shape (straight or cut). The following crosses were carried out.

- Cross 1: Straight-leaved, tall plants were crossed with cut-leaved, dwarf plants. The F_1 generation were all straight-leaved, tall plants.
- Cross 2: The F_1 generation were self-fertilised giving: 899 straight-leaved tall plants, 301 straight-leaved dwarf plants, 290 cut-leaved tall plants, 99 cut-leaved dwarf plants.

'Real-life' ratios will never be perfect

(a) State suitable symbols for the following alleles:
(i) straight-leaved
(ii) cut-leaved
(iii) tall
(iv) dwarf.

(b) (i) Complete the following genetic diagram for the F_1 self fertilisation:

F_1 phenotype: Straight-leaved tall x Straight-leaved tall

F_1 genotype: ______________ ______________

Gametes: ______________ ______________

______________ ______________

(ii) State all the possible genotypes of the F_2 generation.

(c) Suggest a reason why the observed ratio in the F_2 generation is not a precise 9:3:3:1.

3 Coat colour in some hamsters is controlled by two genes on separate chromosomes. The gene for colour can be either B = black or b = white and the gene for coat pattern can either be P = plain or p = mottled.

(a) A black, plain hamster (BbPp) was crossed with a white, plain hamster (bbPp). Complete the following information:

When asked to state genotypes, make sure you have listed all those possible for a phenotype

Parental genotype BbPp bbPp

Gametes ______________

Genotype of offspring ______________

Phenotype of offspring ______________

Ratio of offspring ______________

Test your knowledge

1. ______________ variation gives a normal distribution. This is because it is affected by the ______________ and the environment. It is polygenic, i.e. controlled by several ______________.

2. Frameshift mutations can involve either __________, where a single base is removed, or insertion, where a new __________ is inserted into the sequence. Whereas __________ mutations involve either substitution or inversion.

3. __________ mutations are changes in chromosome structure during synapsis of prophase I and result in a change in the gene sequence. __________ is where a section of one chromosome becomes attached to another and this can lead to Down's syndrome.

4. In genetic engineering, the gene is removed by ____________ ______________, which make staggered cuts in the DNA at specific base pairings, leaving the DNA with '______________' ends. It is placed into a vector, usually a ____________, with DNA ______________.

5. ____________ genes are used to identify transformed bacteria and these are ______________ resistance genes, e.g. ampicillin and tetracyline resistance.

6. Genetically engineered insulin is ______________ to human insulin and causes no ________ ___________ compared to pig's insulin. The disadvantages of genetic engineering include concerns over its ____________ because new pathogens could arise. There are also ____________ issues involved, e.g. whether it is right for people to determine the characteristics of a child.

Answers

1. Continuous, genotype, genes **2** deletion, base, non-frameshift **3** Chromosomal, Translocation **4** restriction endonucleases, sticky, plasmid, ligase **5** Marker, antibiotic **6** identical, side effects, safety, ethical

If you got them all right, skip to page 54

Improve your knowledge

Key points from AS in a Week

Cell division (mitosis and meiosis) pages 35–38

Reproduction pages 61–63

There are two types of variation:

- continuous variation – range of phenotypes controlled by several genes (polygenic) and the environment
- discontinuous variation – distinct types of phenotype in a population controlled by a single gene.

Variations in a population are produced by:

- mutation (point or chromosomal)
- independent assortment of chromosomes in metaphase 1 of meiosis
- crossing over / chiasmata formation during prophase 1 of meiosis
- random fertilisation of gametes.

Morphological and functional changes during evolution are produced by mutations (changes in DNA) occurring during DNA replication.

Point mutations are changes in the sequence of bases in a gene. There are two main types: frameshift and non-frameshift.

In frameshift mutations, the whole sequence (reading frame) is altered, i.e. every single codon (three bases coding for an amino acid) after the mutation is changed, forming a nonsense polypeptide. This can be lethal for a cell, if the mutation occurs prior to mitosis, or fatal to the resultant organism if prior to meiotic division.

The closer to the beginning of the gene, the more severe the effect of mutation

There are two types:

- deletion, where a single base is removed
- insertion, where a new base is inserted into the sequence.

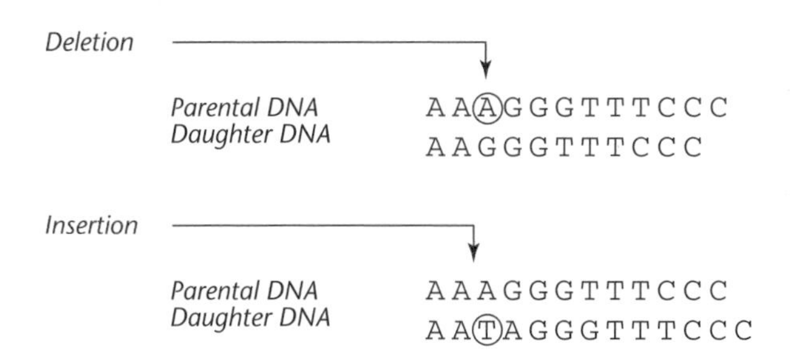

In non-frameshift mutation, the whole reading frame is not altered – i.e. apart from the mutated codon, all other codons remain intact – so only one amino acid is different in the resultant polypeptide. This affects the structure and functioning of the protein only if the amino acid is either involved in the folding of the protein or is located within an enzyme's active site.

There are two types:

- substitution, where a wrong base is substituted for the correct base
- inversion, where two bases in the chain swap places.

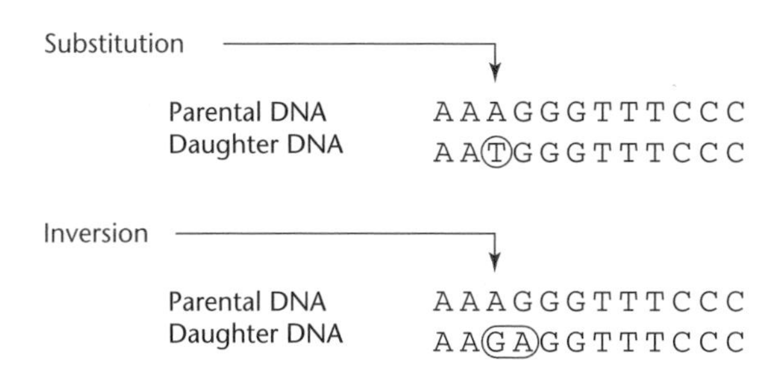

Chromosomal mutations involve changes in the chromosome structure during synapsis of prophase I, when chromatids break and rejoin. This changes the gene sequence. Changes can be:

- deletion – sections of DNA are lost (often fatal)
- inversion – sections within the chromosome are reshuffled
- duplication – a section of DNA is doubled

- translocation – a section of one chromosome becomes attached to another (this can lead to Down's syndrome).

4 Gene technology has been used to manipulate DNA to produce human proteins. This process is called genetic engineering. The mRNA that codes for the protein is identified and is extracted using high density ultracentrifugation. It is converted to copy DNA (cDNA) using an enzyme called reverse transcriptase. It is converted to double-stranded DNA using DNA polymerase. Restriction endonucleases cut the DNA at specific base sequences and extract the gene with 'sticky ends'. It is also used to cut open a vector (plasmid or virus), to ensure that complementary base pairs form. The gene is combined with the plasmid using an enzyme called DNA ligase.

Do not confuse genetic engineering with selective breeding

The vector is inserted into a host cell, usually *E.coli*, by using a cold 0.1 M $CaCl_2$ solution. This causes the recombinant plasmids (plasmids with inserted genes) to adhere to the bacterial cell walls. It is warmed to 37 °C, which encourages the bacterial cells to take up the plasmids and the protein is produced.

5 Transformed bacteria can be identified using marker genes. For example, antibiotic resistant genes can be used (i.e. ampicillin and tetracycline resistance). Within the tetracycline gene is an ECO_{r1} recognition site. If the cDNA has been inserted, this gene is inactivated and the bacteria will only show resistance to the antibiotic, ampicillin. This is easily detected by growing the bacteria on nutrient agar plates, one containing ampicillin and the other containing tetracycline. The transformed bacteria are grown and rapid replication of the plasmids occurs. The protein is synthesised and is extracted from the medium.

6 Social, ethical and economic implications of genetic engineering

Advantages:

- Insulin produced was genetically identical and no side-effects occurred.
- Insulin was originally extracted from animals, which was inhumane and its use conflicted with religious beliefs.
- Chymosin is produced from genetically modified yeast cells and coagulates casein in the dairy industry. Originally, rennin was used, which was obtained from a calf's stomach, and vegetarians refused to eat dairy products that contained rennin.
- Potentially, plants could be genetically modified to be resistant to diseases or have the gene for nitrogen fixation introduced. This would result in an increased crop yield, i.e. be more profitable.
- Production of human hormones, e.g. somatotrophin (human growth hormone) was in short supply but large amounts can be produced by genetic engineering.
- Gene therapy – viruses inserting a healthy gene into cells of individuals suffering from a genetic disorder – could help many people, e.g. those with cystic fibrosis.

Learn these controversial arguments

Disadvantages:

- Safety – genetically modified organisms could have a catastrophic effect on other species, or new pathogens may arise.
- Ethical issues – is it right to transfer genes between species, for companies to patent genes, for people to decide genes to determine the characteristics of a child or alter genes in sperm and ova?

Variation

Use your knowledge

1 Name two processes which take place in meiosis and how they contribute to variation.

Hint: Remember the process of meiosis

2 The diagram below shows the sequences of bases on the part of DNA which codes for haemoglobin.

CACTAGAAC

(a) Draw a sequence of bases after an addition point mutation.
(b) Draw a sequence of bases after a deletion point mutation.
(c) Give the name of a disease caused by a point mutation in the DNA sequence coding for haemoglobin.

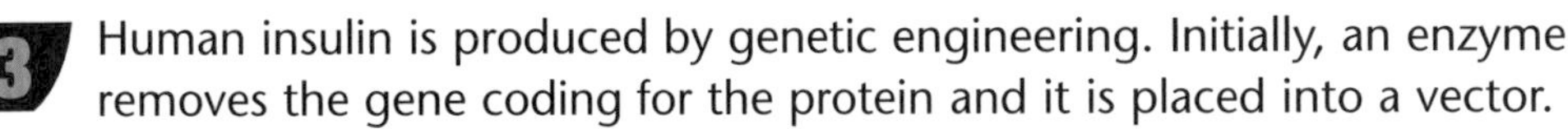

3 Human insulin is produced by genetic engineering. Initially, an enzyme removes the gene coding for the protein and it is placed into a vector.
(a) Name the enzyme that removes the gene.
(b) Name an example of a vector.
(c) Name the enzyme that adheres the gene into the vector.
(d) Why are bacteria used to synthesise the protein?

4 The histogram below shows the height of sunflower plants in a field.

Hint: What factors are involved?

(a) Describe what type of variation is shown. Explain your answer.
(b) Why does the height vary?

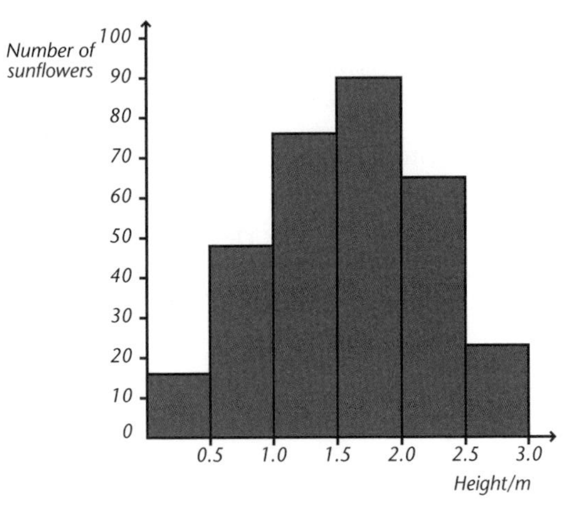

Test your knowledge

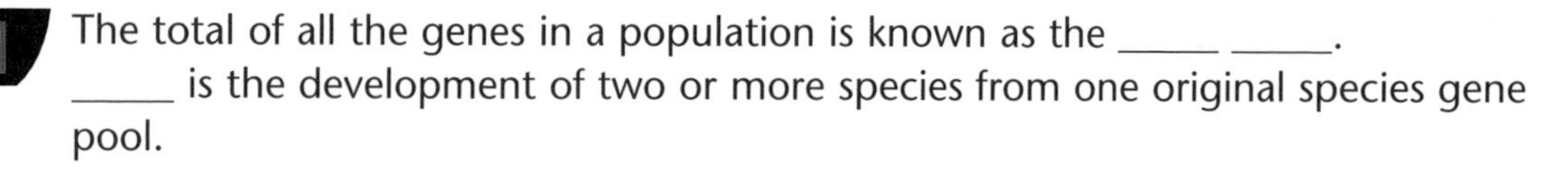

1. The total of all the genes in a population is known as the _____ _____. _____ is the development of two or more species from one original species gene pool.

2. One of the main mechanisms of evolutionary change is _____ selection. If offspring receive _____ characteristics, their chances of survival to reproduce and pass on the _____ are increased.

3. _____ variation is controlled by a single gene. Whereas continuous variation, e.g. height, is controlled by several _____ and the _____.

4. _____ speciation may occur through spatial separation of populations, commonly through _____ isolation, where populations become separated due to geographical features. This prevents exchange of _____ between demes.

5. _____ speciation occurs without geographical isolation. It occurs through two main mechanisms: _____ where individuals contain additional sets of chromosomes, and _____ isolation.

6. _____ are organisms lacking nuclei and other membrane-bound organelles. This kingdom differs from _____ , e.g. amoeba which have envelope-bound nuclei and are often _____ or assembled from similar cells.

7. Taxonomy is the process of ordering groups of organisms sharing common features, called _____. For example, humans are in the _____ chordata and class _____.

Answers

1 gene pool, Speciation **2** natural, advantageous, characteristics/alleles
3 Discontinuous, genes, environment **4** Allopatric, geographical, genes
5 Sympatric, polyploidy, ecological **6** Prokaryotae, protoctista, unicellular
7 taxa, phylum, mammalia

If you got them all right, skip to page 60

Improve your knowledge

1 The gene pool is the total of all the genes in a population. For example, if there are two alleles for wing length (long and vestigial) in a population of 1000 fruit flies, the size of the gene pool for this characteristic is 2000 (2 x 1000). Population genetics studies changes in the gene pool of a population, which may result in evolution.

Key points from AS in a Week
Ecology
pages 67–72

Evolution can be defined in two ways:

- changes in the characteristics of a population over many generations, allowing adaptation to their environment
- development of two or more species from one original species, known as speciation.

Charles Darwin proposed the theory of natural selection as the mechanism for evolutionary change. The modern form (neo-Darwinian theory) can be summarised as follows:

- Ability to survive depends on the specific characteristics of an organism, e.g. strength and speed.
- All characteristics vary within a species. Some variations are advantageous, e.g. faster running to escape predators, and some disadvantageous, e.g. markings that increase visibility to predators.
- Populations produce more offspring than can survive. If offspring receive advantageous characteristics from their parents, their chances of survival to reproduce and pass on the characteristics are increased. In this way, the frequency of an advantageous allele within a population will increase. This leads to a change in the characteristics of a species over time.

Similarly, disadvantageous alleles will become less frequent within a population

Evolution

There are two types of variation in characteristics:

- Discontinuous – distinct types of phenotype in a population, controlled by a single gene, e.g. eye colour.
- Continuous – range of phenotypes, controlled by several genes and the environment, e.g. height. This can be acted upon by natural selection in three ways:
  1. Stabilising selection – selection against extremes, e.g. the optimum weight for human babies is 3.6 kg. Babies born above or below this weight have increased mortality rates.
  2. Directional selection – changes in phenotype in a particular direction, e.g. the fastest cheetah will catch the most prey, so increased speed will be selected for in the cheetah population.
  3. Disruptive selection – selection against intermediate phenotypes, favouring extremes.

A species is defined as individuals capable of interbreeding to produce fertile offspring

4

When populations become geographically isolated, e.g. by rivers or mountains, these demes (isolated populations) are exposed to different environmental conditions, e.g. food availability or climate. This results in natural selection favouring different phenotypes in the different areas, causing the demes to become increasingly genetically different. Eventually, the demes become so genetically diverse that they are reproductively isolated – unable to interbreed to produce fertile offspring. This is known as allopatric speciation.

Sympatric speciation occurs without geographical isolation. Demes found in the same area can separate into reproductively isolated groups through two main mechanisms:

Polyploidy only occurs in plants

- Polyploidy – A process producing individuals containing additional sets of chromosomes, as shown with modern wheat overleaf. This can occur when related species interbreed, or homologous pairs of chromosomes fail to separate in meiosis, giving diploid gametes.

Eincorn wheat (AA) *2n=14*	Wildgrass (BB) *2n=14*	
↓		
Infertile hybrid A (AB) *14 chromosomes*		Infertile because no homologous pairs of chromosomes ∴ no meiosis to make gametes
↓		
Emmer wheat (AABB) *2n=28*		Chromosome number doubles due to DNA replication not followed by cell division
Emmer wheat (AABB) *2n=28*	Goat grass (CC) *2n=14*	
↓		
Infertile hybrid B (ABC) *21 chromosomes*		Infertile because uneven number and no homologous pairs of chromosomes
↓		
Triticum aestivum (AABBCC) *2n=42*		Chromosome number doubles allowing hybrid to become fertile

- Ecological isolation – If a favourable new adaptation arises in a species, individuals possessing it are likely to become reproductively isolated from the rest of the population. For example, they may become separated by their feeding behaviour, although inhabiting the same area.

6 Classification is a system used to help understand the diversity of living organisms, by placing organisms with similar characteristics into groups for easy reference. The binomial system of nomenclature is used, with each organism given a two-part Latin based name, e.g. *Lumbricus terrestris* (earthworm) or *Primula vulgaris* (common primrose). The first part (generic name) states the genus or group of closely related organisms. The second part (specific name) states the species. This is unique to the particular organism.

'Nomenclature' means naming system

The five kingdom system is the current scheme of classification and recognises the kingdoms listed below:

Kingdom	Characteristics	Examples
Prokaryotae	Lack nuclei and other membrane-bound organelles; DNA is circular	*Escherichia coli* Anabena
Protoctista	Eukaryotic organisms with envelope-bound nuclei neither fungi/plants/animals; often unicellular or assembled from similar cells	Amoeba Green algae
Fungi	Non-photosynthetic eukaryotic organisms; non-cellulose cell wall; reproduce by forming spores; lack cilia and flagella	*Mucor* (bread mould) *Agaricus* (mushroom)
Plantae	Multi-cellular photosynthetic eukaryotic organisms; life cycle includes a diploid embryo stage	Mosses Flowering plants
Animalia	Multi-cellular heterotrophic organisms; nervous coordination	Annelids Chordates

7 Taxonomy is the process of ordering groups of organisms called taxa into hierarchies, which attempt to account for their evolutionary origins. The taxa are groups of organisms sharing common features. There are seven levels, with 'kingdom' the highest and 'species' the lowest, as indicated below with the classification of a species of seaweed (bladderwrack) and humans.

Taxon	Description	Bladderwrack *Fucus vesiculasus*	Human *Homo sapiens*
Kingdom	Largest group of organisms with common features	Protoctista	Animalia
Phylum	Major subdivision of kingdom	Phaeophyta	Chordata
Class	Group of related orders	Phaeophyceae	Mammalia
Order	Group of related families	Fucales	Primates
Family	Group of closely related genera	Fucaceae	Hominidae
Genus	Group of related species	*Fucus*	*Homo*
Species	Group of organisms capable of interbreeding to produce fertile offspring	*vesiculosus*	*sapiens*

your knowledge

Hint

1 Closely related species of the dog family are found distributed around the globe, all of which are considered to have evolved from a common ancestor. For example, dingoes are found throughout Australia, coyote along the west coast of North America and jackals in southern Africa.

(a) State the mechanism of isolation most likely to have produced this variety of species.

(b) Explain how these distinct species may have evolved from a common ancestor.

(c) (i) What is the difference between allopatric and sympatric speciation?

(ii) State two examples of sympatric speciation.

Habitat conditions will differ between Australia, America and Africa

2 Following their introduction into Australia in 1859, rabbits spread rapidly until they covered the entire continent by 1905. In an attempt to control the population the disease myxomatosis was introduced. This is a viral disease which passes between rabbits via rabbit fleas. It caused blindness and death in 99.9% of the population, but over the next ten years the rabbit numbers recovered to pre-introduction level.

(a) Explain why the rabbits were able to rapidly colonise Australia.

(b) Suggest two reasons for the large percentage mortality in the rabbit population.

(c) Suggest, with reference to the rabbit gene pool, why not all the rabbits were killed.

Rabbit populations must provide ideal conditions for the spread of the virus

Test your knowledge

1. A __________ is a group of individuals in an area which can interbreed. They are found in __________ , which are communities of organisms interacting with their __________ environment.

2. Populations can increase due to __________ or births. __________ growth occurs when a species colonises a habitat and nothing limits growth.

3. Density-__________ mortality factors such as disease mean the proportion of a population dying increases as density increases.

4. The ____________ factor pH can be tested using a pH _________ or a soil testing kit using an ______________ .

5. A _______________ transect is an ecological technique used to show the distribution of a plant species over an area of ___________ . The numbers of plants are recorded using a ___________ quadrat.

6. A ____________________ of woodlice was estimated using the Lincoln Index. 52 woodlice were caught and marked and after 24 hours, 75 woodlice were caught, 12 of which were marked. The estimated population was ___________ .

7. In a stable ecosystem, the populations are determined by ____________ factors and the diversity of organisms is __________ .

8. Succession from bare ground is called __________ succession. As the soil develops, smaller plants are replaced with larger plants, until the __________ __________ is reached.

Answers

1 population, ecosystems, physical **2** immigration, Exponential **3** dependent
4 abiotic, meter, indicator **5** belt/line, change, frame **6** population, 325
7 biotic, high **8** primary, climax community

If you got them all right, skip to page 66

Improve your knowledge

Key points from AS in a Week
Ecology pages 67–72

1 A population is a number of organisms of the same species living in the same area at a particular time.

A community is a group of populations held together in an area by feeding or competitive interactions.

An environment is the abiotic (non-living) component of an ecosystem, e.g. light intensity, temperature and pH.

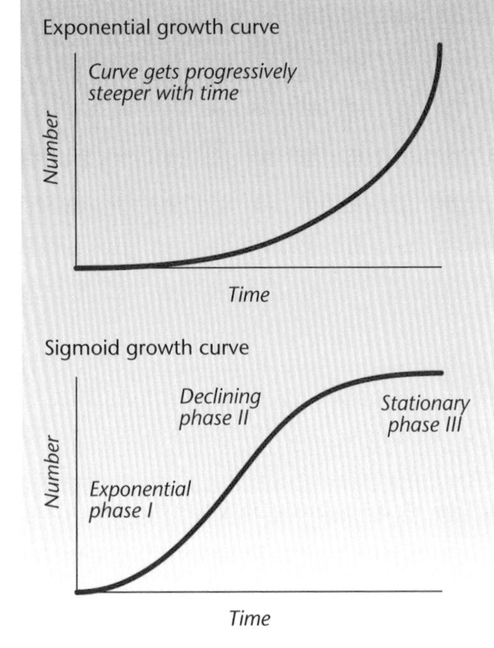

2 If a population is not changing, the growth rate (change in numbers over time) is zero. Numbers in a population can increase (positive growth rate), due to births and immigration, or decrease (negative growth rate), due to deaths and emigration.

Population is growing at its maximum rate, with births higher than deaths. This occurs when a species colonises a new habitat and no factors, e.g. food, space, light, limit the population growth.

Exponential growth cannot continue. Population number will reach a level where there are too many individuals and intraspecific competition (competition between individuals of the same species) occurs for resources, so growth rate declines.

All habitats have a maximum population size that can be supported

3 Competition is a major factor determining population size. A sigmoid growth curve shows that intraspecific competition is a density-dependent mortality factor. The proportion dying increases as population density increases because the level of competition for resources, e.g. food, space, water, light, increases. Other examples of density-dependent factors are predation and contagious diseases.

Density-dependent means percentage dying increases as density increases

Climate is a density-independent mortality factor, with the proportion of population dying independent of the density, e.g. a cold winter will kill individuals of a young bird population, whatever the population density.

Abiotic factor	Measurement technique
pH	pH meter/soil testing kit with indicator
Oxygen content of water	Oxygen probe
Light intensity	Light meter

Ecological techniques can be used to investigate the numbers and distribution of organisms. A plant population can be estimated using a frame quadrat. An area is defined using a table of random numbers to provide coordinates. The $0.25\,m^2$ quadrat is placed at the coordinate and the presence or absence of a species is recorded. This technique is repeated and the percentage frequency is calculated as:

$$\frac{\text{Number of quadrats in which the species was present}}{\text{Total number of quadrats}} \times 100$$

A line or belt transect can be used to show the distribution of a plant species. A line is placed across an area of change (e.g. rocky shore) and the number of plant species is recorded.

The limitations of these methods include:

- Do not take into account the clumping of organisms.
- Some areas are left unsampled.
- The same area could be sampled more than once.
- Do not take into account seasonal variation.

6 The number of mobile organisms in a population can be calculated by the Lincoln index, also known as the mark-release-recapture method.

The organism is initially trapped (e.g. in a pitfall trap), marked using an insoluble paint and released. Another sample is captured 24 hours later in the same area and the number of marked and unmarked organisms is recorded. An estimate of the population can be calculated using the following equation:

$$\text{Estimated size of the population} = \frac{n_1 \times n_2}{n_3}$$

Practise using this formula

where

n_1 = number of organisms captured, marked and released

n_2 = total number of organisms captured in the second sample

n_3 = number of marked individuals recaptured

This assumes that no emigration or immigration or births or deaths have occurred. Also, it assumes random mixing of the marked animals in a population and that the marking does not poison the animal or increase predation.

In an unstable ecosystem, populations are determined by abiotic factors and the diversity of organisms is usually low. In a stable environment, the environment is less hostile and the populations are determined by biotic factors. The diversity of organisms is usually high and can be calculated from the formula:

$$D = \frac{N\,(N-1)}{\Sigma\, n\,(n-1)}$$

where D = diversity index, N = total number of plants, n = number of individuals per species, Σ = sum of

This is commonly used for measuring plant diversity. A sample area is defined and each species is identified and counted. This can be seen in succession.

Succession is the development of a stable community (the climax community) over time, through a number of stages. This results in a long-term change in the composition of the community due to the modifying effect that organisms have on their environment.

Succession from bare ground is known as primary succession and can take hundreds of years. Secondary succession occurs where vegetation has already been present, e.g. burnt heathland or cleared woodland, and occurs in a shorter period of time.

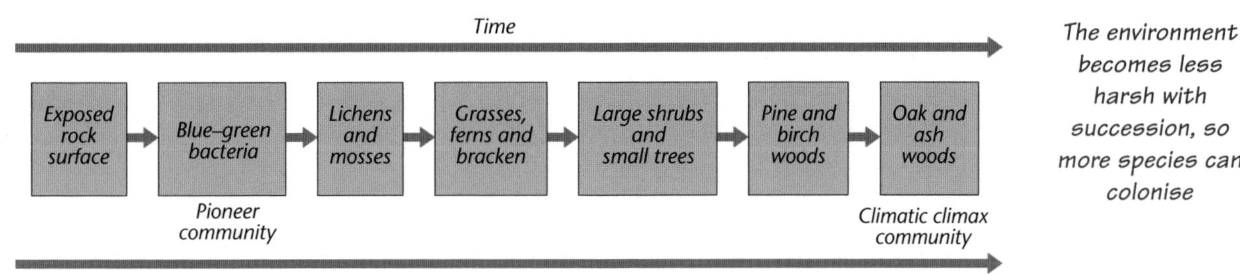

The environment becomes less harsh with succession, so more species can colonise

- Algae and lichen form the pioneer community and are able to colonise areas with no organic matter, e.g. bare rock from a landslide. They cause the accumulation of organic material and weathered rock, forming a young soil.
- The soil allows mosses, ferns and small herbaceous plants to grow. These species out-compete and replace the pioneer-community species (interspecific competition), and more soil accumulates.
- As the soil develops further, interspecific competition results in these small plants being replaced by larger plants, until the climax community is reached. This community is in equilibrium with the climate. In England the climax community is usually deciduous woodland.

Use your knowledge

Hint

1 Define the following ecological terms:
(a) community
(b) ecosystem.

Remember the definitions

2 A field has been burnt and left bare.
(a) Describe the changes that will take place over the next ten years.
(b) What usual agricultural practices take place which prevent succession?

3 (a) Define a population.
(b) Define intraspecific competition.
(c) Suggest three factors that may influence the length of time taken for a wild population to reach the carrying capacity.

4 The graph shows the population change over 24 hours in yeast cells following introduction into a large volume of well-aerated nutrient solution.

What factors affect population?

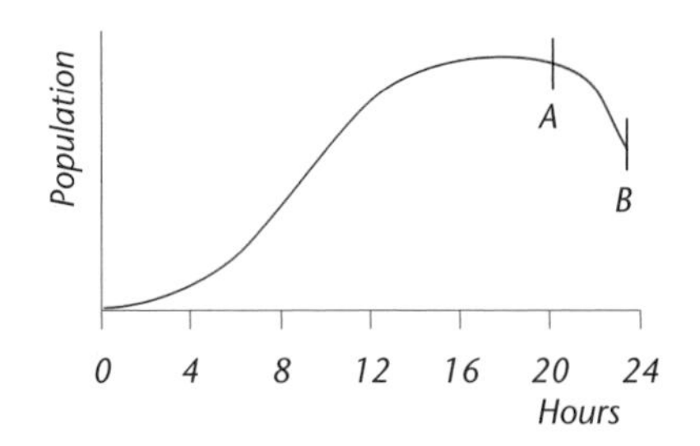

(a) Suggest why the population changes between points A and B.
(b) Suggest two factors that may determine the population's carrying capacity.

5 An estimate of a mobile population can be calculated using the Lincoln index.
(a) Name two methods of marking the animals.
(b) What assumptions have been made?

Remember the equation to calculate the Lincoln index

Nutrient Cycles and Pollution

Test your knowledge

1 _____ is the conversion of nitrates to nitrites and then to nitrogen gas. Nitrogen fixation converts atmospheric nitrogen into _____ , using the enzyme _____ .

2 The enrichment of a watercourse by _____ or phosphorus fertilisers is termed _____. Decomposition of the algal bloom produced by _____ bacteria reduces the water's dissolved oxygen level.

3 All plants and animals release CO_2 via _____ , and photosynthetic organisms take up carbon in the form of _____ . Precipitation removes atmospheric CO_2 into _____ rocks.

4 Greenhouse gases allow entry of _____-_____ solar radiation, and prevent loss of _____-_____ radiation, _____ from the Earth's surface.

5 _____ is the clearance of woodland, often to create agriculture land. The consequences include increased _____ erosion, _____ of rivers and habitat loss.

6 Desertification can be caused by over-_____ , when poor quality land is used for crop production. _____ declines, reducing crop growth, exposing the soil to wind and _____ .

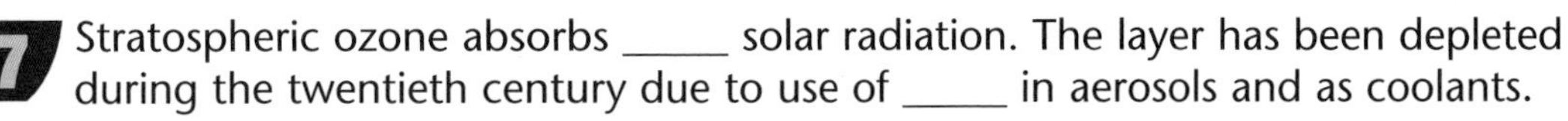

7 Stratospheric ozone absorbs _____ solar radiation. The layer has been depleted during the twentieth century due to use of _____ in aerosols and as coolants.

8 Acidic _____ is the deposition of acidic gases from the atmosphere, produced by combustion of _____ fuels. The consequences include _____ of freshwater.

9 Untreated sewage pollution contains _____ matter, which when broken down by _____ bacteria reduces dissolved oxygen levels through increasing the _____ _____ demand.

Answers

1 Denitrification, ammonia, nitrogenase **2** nitrogen, eutrophication, aerobic
3 respiration, CO_2, carbonate **4** short-wave, long-wave, re-radiated
5 Deforestation, soil, silting **6** cultivation, Fertility, rain
7 shortwave/incoming, CFCs **8** precipitation, fossil, acidification
9 organic, aerobic, biochemical oxygen

If you got them all right, skip to page 73

Nutrient Cycles and Pollution

Improve your knowledge

1 Organisms require a source of nitrogen because it is a component of many organic molecules, e.g. nucleic acids and proteins. They obtain nitrogen from various sources, as illustrated in the nitrogen cycle below:

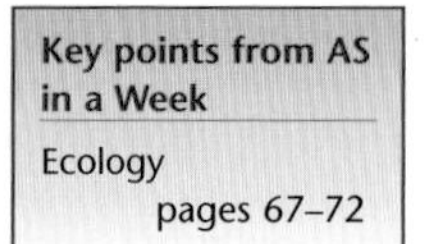

Identify the three main processes in the nitrogen cycle

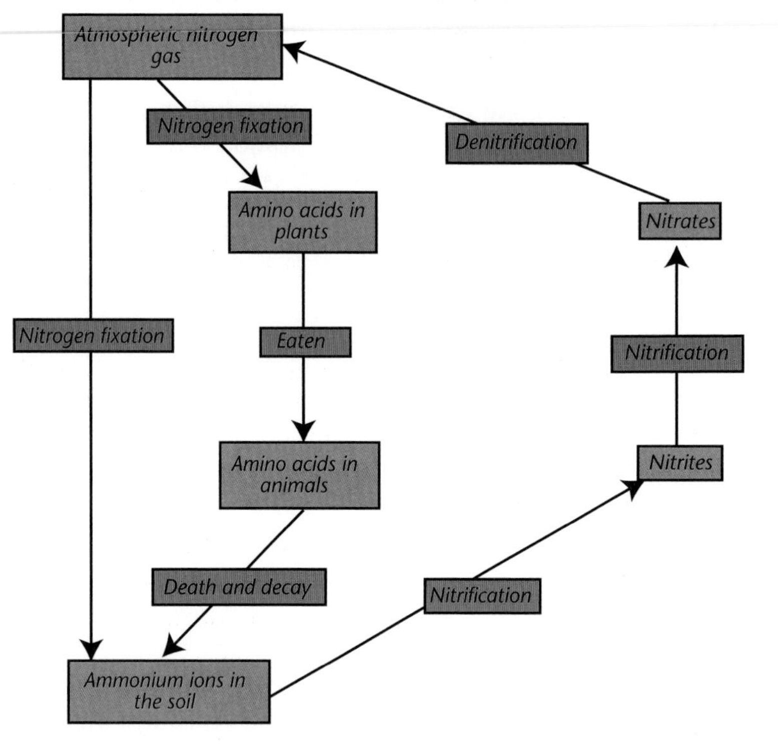

- Decay or decomposition is the conversion of nitrogen-containing compounds in dead plants and animals, faeces and urine into ammonium ions by bacteria and saprophytic fungi.
- Nitrification is the conversion of the ammonium ions to nitrites, then to nitrates, by nitrifying bacteria.

	Nitrosomonas		*Nitrobacter*	
NH_4^+ Ammonium	→	NO_2^- Nitrite	→	NO_3^- Nitrate

Nutrient Cycles and Pollution

- Denitrification is the conversion of nitrates to nitrites and then to nitrogen gas, which is lost from the soil into the atmosphere. This occurs by denitrifying bacteria, e.g. *Pseudomonas denitrificans,* under anaerobic conditions (such as water-logged soils).
- Nitrogen fixation is the conversion of atmospheric nitrogen into ammonia, using the enzyme nitrogenase. Nitrogen-fixing bacteria can be found free living in the soil, e.g. *Azotobacter*, or in root nodules of leguminous plants (such as clover and peas), e.g. *Rhizobium*. These bacteria receive carbohydrate from the plant and in return provide ammonium ions.

2 Fertilisers are added to soil to make up for lost nutrients. Additions of nitrogen or phosphorus fertilisers can damage ecosystems. The fertiliser may leach from the soil and enter rivers and lakes. This enrichment of water is termed eutrophication.

Eutrophication significantly increases the growth of aquatic plants, causing 'algal blooms'. The death and subsequent decomposition of these plants by aerobic bacteria reduces dissolved oxygen level. The water has an increased biochemical oxygen demand, and this deoxygenation reduces the species diversity.

Leaching is the loss from soil in solution

3 The diagram below shows the carbon cycle. CO_2 is released via respiration in all organisms, and photosynthetic organisms take up carbon in the form of CO_2. Carbon is removed from the cycle into long-term store by rock formation, e.g. coal or calcium carbonate (limestone).

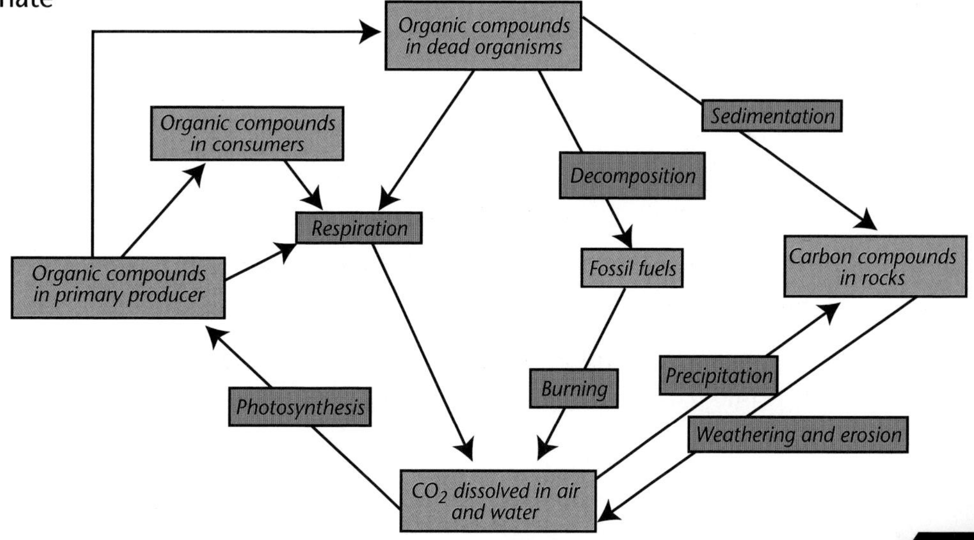

Nutrient Cycles and Pollution

4 The greenhouse effect is the natural warming of the Earth's atmosphere due to greenhouse gases, e.g. CO_2 and O_3. These gases allow entry of shortwave radiation from the sun, but delay the loss of re-radiated longwave radiation, causing an overall warming effect.

A change in climate will cause a change in species diversity

Human activities have increased the concentration of greenhouse gases in the atmosphere, i.e. combustion of fossil fuels in power stations or motor vehicles. This enhanced greenhouse effect has been linked with global climate change, including increased temperatures and declining rainfall.

5 Deforestation is the destruction of forests and woodland at a rate which exceeds replanting or natural regeneration. The causes of deforestation include:

- demand for timber
- clearing land for agriculture, e.g. cattle rearing or arable cropping
- clearing land for roads, housing and industry
- demand for firewood
- damage by pollution, e.g. acid precipitation.

Deforestation has significant consequences on the environment:

- Increased atmospheric CO_2 concentration due to reduced uptake via photosynthesis.
- Habitat destruction and reduction in species diversity, due to the large numbers of species that forests can support.
- Disruption to the hydrological cycle. Removal of trees reduces cloud formation and rainfall, as transpiration is reduced. Trees also bind soil together and take up large volumes of water. When they are removed, low-lying areas are prone to flooding.
- Reduced soil fertility as the increased water flow washes nutrients out of the soil and erodes the top soil.
- Atmospheric pollution can result from the clearing of trees by burning.

Desertification is the conversion of fertile land into infertile land. The causes of desertification include:

- Over-cultivation. Increasing human population has resulted in increasing pressure on agricultural land. For example, poor quality land is now used for crop production and existing land is more intensively farmed. Over time, fertility declines on this land as nutrients are not replaced. This reduces crop growth, exposing the soil to wind and rain, eroding the top soil and leaching out nutrients.
- Over-grazing. Increasing crop growth results in cattle being concentrated into smaller areas. This reduces vegetation growth and bare ground increases, exposing the soil to wind and rain erosion. Cattle also physically erode the land through trampling.
- Poor irrigation. Irrigation systems use a dilute salt solution. As the water evaporates, soil salt concentration increases. This can be toxic to plants, causing exposure and erosion of the soil.
- Over-water extraction. Dam building, or over-extraction from rivers or lakes, reduces available water downstream, with the surrounding land becoming prone to desertification.
- Climate change. Global warming and reduced rainfall cause desertification in vulnerable areas.

Desertification will significantly reduce species diversity

7

Stratospheric ozone (O_3) found at an altitude of between 15 km and 30 km, absorbs short-wave radiation from the sun, protecting living organisms from this radiation that can cause skin cancers and DNA mutation.

Human activity has caused significant thinning of the O_3 layer, particularly at the poles, mainly due to the use of CFCs (in aerosols and as coolants) and nitrous oxides (released from artificial fertilisers). These pollutants, when released into the atmosphere, break down O_3 faster than it is reformed. This has led to increasing cases of skin cancer and reduced crop yield due to damage to vegetation, particularly in the southern hemisphere where greatest thinning has occurred.

Nutrient Cycles and Pollution

Acidic precipitation is the deposition of acidic gases (pH 4 – 4.5) from the atmosphere, produced by combustion of coal and oil (fossil fuels). The principal gases are sulphur and nitrogen dioxide.

Acidic precipitation is a trans-boundary pollutant, moving between countries

Gases may be deposited directly onto surfaces (dry deposition) or dissolved in water droplets (wet deposition), causing varied effects:

- decline in forests due to cell damage by acidic soil conditions and by making toxic heavy metal ions in the soil available to be taken up by roots
- acidification of freshwater, killing fish and invertebrates
- chemical weathering of buildings and monuments.

9

Water pollution occurs in a diversity of different forms:

- Sewage – Untreated sewage contains organic matter, e.g. faeces, which is broken down in water by large populations of aerobic bacteria. This increases the biochemical oxygen demand, so the dissolved oxygen levels fall, reducing the number of fish and aerobic organisms in the water. Sewage may also contain toxic chemicals and pathogens.
- Thermal pollution – Factories and power stations release warm water into watercourses. Increased temperature reduces the solubility of oxygen in water, so dissolved oxygen levels decrease. In addition, certain species' behaviour, e.g. migration and spawning times, may be affected.
- Oil pollution – Oil enters into watercourses from tanker spillage and small discharge from industries and refineries. This oil may clog the feathers of sea birds, or contain toxic substances that can cause the death of marine organisms. Oil washed ashore can cause the death of plants and sedentary organisms, e.g. limpets.

Human activity in an ecosystem often causes a decline in species diversity

Nutrient Cycles and Pollution

20 minutes

Use your knowledge

Hint

1 Suggest an explanation for the following statements:

(a) Crops grown on soils following deforestation may fail within the first five years of cultivation.

(b) Nitrogen fertiliser applied to poorly drained soils is unlikely to have a significant effect on crop productivity.

How do soluble nutrient ions enter the soil?

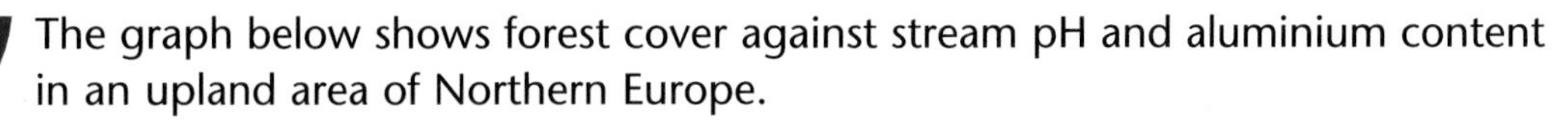

2 The graph below shows forest cover against stream pH and aluminium content in an upland area of Northern Europe.

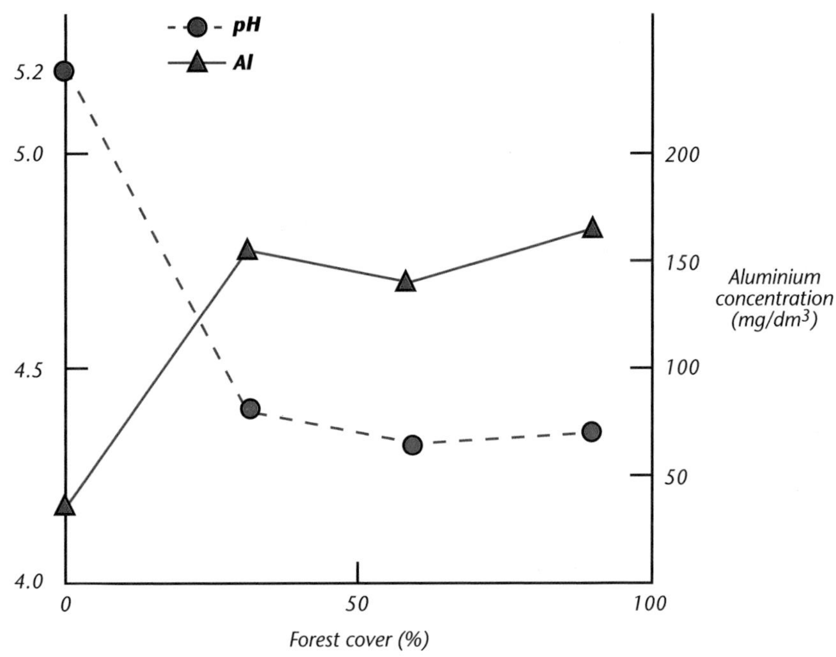

(a) Describe the relationship between forest cover and stream pH.

(b) Suggest an explanation for the trend in aluminium concentrations.

(c) Discuss effects on the stream ecosystem of acidification.

How will acidic conditions affect aluminium ions?

3 (a) Explain what is meant by the term eutrophication.

(b) Suggest explanations for the following observations on eutrophication:

(i) Dissolved oxygen concentration of water decreases.

(ii) Large plants on the bed of the lake or river disappear.

Eutrophication causes growth

Improve your knowledge

There will be a large emphasis in many questions in the A2 examination on testing the synoptic component. This will be marked by the side of a question with the letter 'S'. It will include:

- Drawing graphs – To attain maximum marks, plot the points accurately, draw the axes correctly, ensure that the scale is large (i.e. the graph should be larger than half the graph paper) and join each plotted point with a ruler.
- Experiments – Data will be given and analysis of the results will be expected. In addition, some questions may include details about the method of the investigation. In this case, always include the control variables and methods to ensure accuracy.
- Statistics – Be able to carry out a χ^2 (chi-squared) statistical test. This is used to compare (observed) experimental values with expected values. Using this method, it can be deduced whether the results are significant or insignificant, i.e. have the results occurred due to chance alone? The equation for a χ^2 statistical test is:

$$\chi^2 = \Sigma \frac{(O-E)^2}{E}$$

Σ = 'the sum of'

The number of degrees of freedom is calculated by (number of rows – 1) × (number of columns – 1).

Use tabulate χ^2 values to find out if your results are significant

- Essays – There will be a choice of two essay titles and the composition will be marked as follows: 16 marks for scientific content; 3 marks for breadth of knowledge; 3 marks for relevance; 3 marks for written communication.

It is essential that you begin by planning the essay and that you have thoroughly revised the AS biology material. It is not difficult to attain a high standard in your essay, even though this is the most disliked area of a biology A2 examination.

your knowledge

Hint

A genetics experiment was carried out on Drosophila flies. The results of this dihybrid inheritance experiment for the F_2 generation were:

	Observed	Expected
Black body, vestigial wings	11	16
Grey body, normal wings	178	144
Black body, normal wings	52	48
Grey body, vestigial wings	15	48

(a) Calculate χ^2.

(i) Initially, calculate (O – E) and $(O - E)^2$

(ii) Divide each of the $(O - E)^2$ values by E and add these values together. This is the chi-squared value.

(iii) Calculate the number of degrees of freedom.

(iv) Use a tabulated table of results and if your value is greater than the tabulated value, the results are significant.

Remember the equation to calculate χ^2

2 Write an essay on 'The properties and importance of enzymes'. (25 marks)

Initially write a plan stating biological concepts to be included

Exam Practice Questions

Describe how you would set up an experiment to demonstrate the effect of mineral ions in wheat plants. [6]

Anita carries out an experiment to find out if there is a difference in species diversity between the embryo and semi-fixed dunes.
The results she obtained are shown below:

Embryo dunes

Species	Number
Marram grass	1008
Sea rocket	23
Lyme rocket	4
Seasand wort	7

Semi-fixed dunes

Species	Number
Marram grass	453
Sea spurge	32
Common cat's ear	278
Common ragwort	283
Common groundsel	13
Lesser hawksbit	11
Common dandelion	2
Rosebay willowherb	1
Evening primrose	1

(a) Calculate the species diversity for the embryo and semi-fixed dunes using the following equation: [4]

$$D = \frac{N(N-1)}{\sum n(n-1)}$$

(b) What do the results indicate? [1]

(c) Why does the pH of the soil decrease from the embryo dunes to the semi-fixed dunes? [1]

Exam Practice Questions

3 (a) Throughout Europe, the average height of both males and females has been continually increasing over at least the last century.

(i) Suggest whether this observation is more likely to be due to genetic or environmental factors. [1]

(ii) Explain your answer to (i). [2]

(iii) What environmental factors may influence male and female height? [2]

(b) Define the following terms:

(i) Pure breeding. [1]

(ii) Test cross. [2]

(iii) Allele. [1]

(c) A mule is a hybrid produced by mating a male donkey with a female horse.

(i) Explain why the mule will be infertile. [2]

(ii) Suggest why a farmer may wish to produce a hybrid. [2]

4 (a) Define the term 'biodiversity'. [2]

(b) Modern scientific theory postulates that the temperature of the atmosphere is increasing due the 'enhanced greenhouse effect'.

(i) Explain what is meant by the term 'enhanced greenhouse effect'. [3]

(ii) The 'enhanced greenhouse effect' has been suggested to result in loss of biodiversity. Suggest three reasons why this may have occurred. [3]

(iii) How may levels of atmospheric CO_2 be reduced? [3]

(c) With reference to biodiversity, explain the importance of maintaining river water quality. [5]

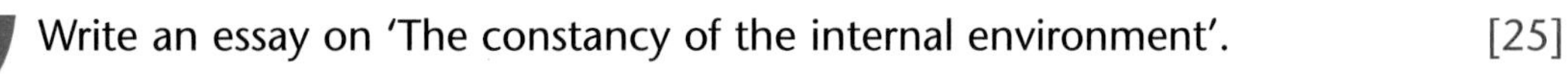

5 Write an essay on 'The constancy of the internal environment'. [25]

Use your Knowledge Answers

Respiration

Fixed mass of sugar (glucose/sucrose); fixed mass of yeast; fixed volume of water; constant temperature using thermostatically controlled water bath; measure the time for the indicator to turn yellow using a stopwatch; compare the end colour with a standard; repeat the readings and calculate an average; nitrogen flow must be kept constant; used a fixed volume of indicator.

(a) Glucose.
(b) Cytoplasm.
(c) NAD/FAD are hydrogen acceptors; electrons transferred along carriers; progressively lower energy levels; energy released forms ATP; oxygen is the final acceptor.

Homeostasis

(a) Urine – greater concentration of urea; water reabsorbed in nephron; glucose absent (present in filtrate); reabsorbed in PCT; greater concentration of soluble ions; water reabsorbed in nephron.
(b) Plasma – contains plasma proteins/absent in filtrate; too large to cross basement membrane; filtrate more dilute/higher water potential; due to absence of proteins.

(a) (i) beta cells in Islets of Langerhans
(ii) insulin
(b) liver
(c) Nervous impulses to the temperature regulating effectors; faster than hormonal control of blood glucose.

(a) H_2O leaves descending limb by osmosis; descending limb impermeable to Na^+/Cl^-; Na^+/Cl^- pumped out of ascending limb by active transport; ascending limb impermeable to H_2O.
(b) H_2O is reabsorbed along with solutes.
(c) ADH present; because concentrated urine produced; water reabsorbed in DCT/collecting duct.
(d) PCT
(e) Rate of ultrafiltration exceeds rate of reabsorption; not all the glucose can be reabsorbed at the PCT.

Use your Knowledge Answers

Nervous coordination

1

(a) A – synaptic knob/membrane
B – cell body
C – node of Ranvier
D – Schwann cell
E – receptor

(b) Insulates membrane; increases speed of transmission.

(a) three neurones drawn correctly (see page 22)

(b) automatic/involuntary; rapid response to a stimulus

(c) muscle/gland

Sensitivity and responses

(a) rods; cones

(b) Rods sensitive to low light intensity; cones sensitive to high light intensity; three types of cones, each sensitive to red, blue, green light; rods sensitive to all colours/wavelengths.

(c) Light-sensitive pigment iodopsin; three different types sensitive to red, green or blue light/wavelengths; specific wavelengths break the specific form of iodopsin causing depolarisation/action potentials; combinations of stimulation used for other colours.

(d) Light refracted by cornea/aqueous humour/vitreous humour; ciliary muscles contract; reducing tension on suspensory ligaments; elastic lens becomes more rounded; more refraction to produce sharp image on retina.

(e) Iris muscles regulating diameter of pupil; in bright light circular muscles contract and radial muscles relax; pupil becomes smaller so less light enters; in dim light circular muscles relax and radial muscles contract; pupil diameter increases so more light enters.

2

(a) Increased speed of movement; increased number of changes in direction per given time; non-directional response.

(b) Increased likelihood of finding favourable conditions; avoid unfavourable conditions; protective function against dessication.

(c) (i) Kinesis – non-directional increase in movement; response to an unfavourable stimulus.

(ii) Taxes – directional response; either towards or away from an unfavourable or favourable stimulus.

Photosynthesis

(a) grana/thylakoid/quantosomes

(b) (i) photolysis

(ii) passed to PS2

(iii) reduce NADP

(a) (i) Plant A; greater range of wavelengths/blue and red both present.

(ii) Plant C; green light is reflected/little is absorbed.

(b) Concentration of CO_2/H_2O availability/humidity/mineral ion availability.

Use your Knowledge Answers

Inheritance

1

(a) (i) Chromosomes identical in size and shape; carrying the same genes at the same loci.
(ii) Gene codes for a characteristic; alleles are different forms of a gene.
(b) (i) white flowers, oval leaves
(ii) white flowers, round leaves
(c) (i) WWRr; WWRR; WwRr; WwRR
(ii) wwRr/wwRR; wwrr

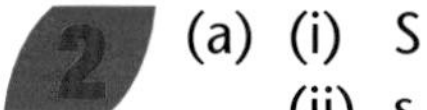

(a) (i) S
(ii) s
(iii) T
(iv) t
(b) (i) SsTt, SsTt, ST, St, sT, st
(ii) SSTT, SSTt; SsTT; SsTt, ssTT, ssTt; SsTt, SStt, sstt
(c) Ratio approximately 9:3:3:1; differences are due to random fertilisation.

(a) BP, Bp, bP, bp; bP, bp; BbPP, BbPp, Bbpp, bbPP, bbPp, bbpp; black plain, black mottled, white plain, white mottled; 3:3:1:1.

Use your Knowledge Answers

Variation

Chiasmata formation/crossing over during prophase I; site of exchange of genetic material; independent assortment during metaphase I; different combinations of genes occur in each gamete.

(a) Any example with 1 base added, e.g. CAACTAGAAC.
(b) Any example with 1 base deleted, e.g. CACAGAAC.
(c) sickle cell anaemia

(a) restriction endonuclease
(b) plasmid/virus
(c) DNA ligase
(d) Replicate every 20 minutes; simple nutritional requirements; occupy little space.

(a) Continuous; normal distribution.
(b) Variation is polygenic and affected by the genotype and environment.

Use your Knowledge Answers

Evolution

(a) geographical isolation

(b) Common ancestor (dogs) evolved; radiated out; isolated in different environments; mutations occur; producing variation; natural selection acts on variation; eventually populations cannot interbreed.

(a) Population characteristics suited to conditions; high reproductive rate; short time to reach sexual maturity; abundant food; absence of predators.

(b) No resistance/immunity in gene pool; dense populations/close association in burrows; disease spread rapidly through population.

(c) Development of resistance/immunity; these individuals survived and bred; allelic frequency of advantageous gene increased in population; increasing percentage of population immune.

Use your Knowledge Answers

Ecology

(a) Community – populations of all species; living in the same habitat.
(b) Ecosystem – community of organisms interacting with their environment.

(a) Pioneer species arrive by seed dispersal; low species diversity initially; plants die and are decomposed to humus; increase in soil nutrients; development of shrubs; growth of trees leading to climax vegetation, e.g. deciduous woodland.
(b) Ploughing; planting new crops; use of herbicides; grazing by sheep.

(a) Group of freely interbreeding members of the same species in a habitat at a given time.
(b) Competition within species for food/light/space, etc.
(c) Time taken to reach sexual maturity; gestation period; number per litter.

(a) Increased competition; build-up of toxic waste.
(b) Food availability; competition; space.

(a) Waterproof paint spot; clip abdominal hairs; tag animals.
(b) No emigration/immigration; no births or deaths; random mixing within a population.

Nutrient cycles and pollution

(a) Nutrients held in trees; soil nutrient content is low; crops rapidly remove remaining nutrients; no decomposition/nutrient recycling from dead plant material.

(b) Poorly drained soils may become anaerobic/lack air spaces containing O_2; denitrifying bacteria may convert nitrates into N_2.

(a) With increasing forest cover, stream pH decreases; qualify with data from graph.

(b) Acid precipitation reduces soil pH; solubility of aluminium increases; aluminium leached into watercourses.

(c) Decreased species diversity; decreased fish stocks; decreased river invertebrates/crustacea; reduced reproductive rates in remaining species.

(a) Excess input/enrichment into watercourses with nutrients; e.g. nitrate/phosphate; from fertilisers/detergents.

(b) (i) Surface algal blooms/increased phytoplankton; reduces light penetration/reaching plants; photosynthesis cannot occur.

(ii) Abundant dead algae/phytoplankton; decomposed/broken down by aerobic bacteria; removes dissolved oxygen/increases biochemical oxygen demand.

Use your Knowledge Answers

Synoptic assessment

1 (a) (i)

	Observed	Expected	(O–E)	$(O–E)^2$
Black vestigial	11	16	–5	25
Grey normal	178	144	34	1156
Black normal	52	48	4	16
Grey vestigial	15	48	–33	1089

(ii) χ^2 : 1.5625 + 8.0277777 + 0.3333333 + 22.6875 = 32.2

(iii) degrees of freedom = (4–1) × (2–1) = 3

(iv) The calculated value is greater than the tabulated value, which is 7.815 at the 5% significance level. Therefore, the results are highly significant (i.e. there is a 5% chance that the results have occurred due to chance).

2 Properties: Tertiary structure protein; specific three-dimensional shape; substrate can bind to the active site forming an enzyme-substrate complex; lock and key/ induced fit theory; competitive/non-competitive inhibition; effect of cofactors, prosthetic groups; catalyst by lowering the activation energy.

Importance: Protein synthesis; anabolic/catabolic reactions; effect of mutations; digestion; metabolism; intra-cellular and extra-cellular enzymes; any other suitable examples.

Exam Practice Answers

1 Set up four boiling tubes with young wheat plants of a similar age; wrap aluminium foil around the roots to prevent the growth of algae; in one boiling tube, add all the mineral ions to act as a control; in the remaining boiling tubes, add all mineral ions except one; allow them to grow for one week with the same light intensity; and bubble oxygen at the same rate into the solution for aerobic respiration; record the deficiency symptoms.

(a) embryo dunes

$$D = \frac{1042\ (1041)}{(1008 \times 1007) + (23\times22) + (4\times3) + (7\times6)}$$

$= 1.07$ (3.s.f.)

semi-fixed dunes

$$D = \frac{1074\ (1073)}{(453\times452)+(32\times31)+(278\times277)+(283\times282)+(13\times12)+(11\times10)}$$

$= 3.18$ (3.s.f)

(b) The species diversity is much greater on the semi-fixed dunes compared to the embryo dunes.

(c) The soil is more acidic in the semi-fixed dunes compared to the embryo dunes because plant material has decomposed to form humus.

3 (a) (i) Environmental.
(ii) Gene pool unlikely to change in 100 years; little relevant selection pressure.
(iii) nutrition; disease

(b) (i) Homologous for a particular characteristic.
(ii) Crossing an individual of unknown genotype with a homozygous recessive; to determine the genotype.
(iii) Alternative forms of a gene.

(c) (i) Chromosomes cannot pair up in prophase I; unpaired chromosomes may not enter gamete.
(ii) Hybrid vigour; producing greater strength/ability to do more work.

Exam Practice Answers

4

(a) Number of different species in an area; at a particular time/eq.

(b) (i) greenhouse effect is the natural warming of the atmosphere; due to absorption of longwave/re-radiated radiation; by CO_2/NO_2/CH_4/CFC/O_3; enhanced due to increased concentration of greenhouse gases through human activity.

(ii) Habitat loss due to rising sea level/flooding; due to increased desertification; evolution too slow for adaptations to changing climatic conditions; invasion/out-competing from immigrating species.

(iii) reforestation; afforestation; reduce fossil fuel combustion; reduce deforestation; reduce emission from power stations.

(c) Quality can be reduced by sewage/fertilisers/thermal pollution/industrial waste/oil; biodiversity decreases due to reduced number of species able to cope with these conditions; bacterial populations increase which may be pathogens/produce offensive smells; reduced recreational usage of water; increased treatment cost for drinking water.

'Constancy of an internal environment'

Definition; importance of maintaining constancy; involvement of negative feedback; factors to be controlled, e.g. glucose concentration, water, temperature, carbon dioxide and oxygen concentration; involvement of the hypothalamus; and hormones, e.g. insulin and glucagon; ADH secretion to increase permeability of distal convoluted tubule and collecting ducts; medulla oblongata – involvement in respiratory rate; stimulating external intercostal muscles to contract; increasing O_2 input and CO_2 output; thermoregulation; physiological mechanisms to increase/decrease body temperature; sweating/vasoconstriction/vasodilation/pili erector muscles/behavioural mechanisms.